Genetics for ENT Specialists

The Remedica Genetics for... Series
Genetics for Cardiologists
Genetics for Dermatologists
Genetics for Endocrinologists
Genetics for ENT Specialists
Genetics for Hematologists
Genetics for Oncologists
Genetics for Ophthalmologists
Genetics for Orthopedic Surgeons
Genetics for Pediatricians
Genetics for Pulmonologists
Genetics for Rheumatologists
Genetics for Surgeons

Published by Remedica

32–38 Osnaburgh Street, London, NW1 3ND, UK

20 North Wacker Drive, Suite 1642, Chicago, IL 60606, USA

Email: info@remedicabooks.com

www.remedicabooks.com

Publisher: Andrew Ward

In-house editor: Cath Harris

Design and Artwork: AS&K Skylight

Remedica is a member of the AS&K Media Partnership

ISBN 1 901346 64 1

ISSN 1472 4618

British Library Cataloguing in-Publication Data

A catalogue record for this book is available from the British Library

Genetics for ENT Specialists

Dirk Kunst, Hannie Kremer, and Cor Cremers
University Medical Centre Nijmegen,
The Netherlands

Series Editor
Eli Hatchwell
Investigator
Cold Spring Harbor Laboratory

Introduction to the Genetics for... series

Medicine is changing. The revolution in molecular genetics has fundamentally altered our notions of disease etiology and classification, and promises novel therapeutic interventions. Standard diagnostic approaches to disease focused entirely on clinical features and relatively crude clinical diagnostic tests. Little account was traditionally taken of possible familial influences in disease.

The rapidity of the genetics revolution has left many physicians behind, particularly those whose medical education largely preceded its birth. Even for those who might have been aware of molecular genetics and its possible impact, the field was often viewed as highly specialist and not necessarily relevant to everyday clinical practice. Furthermore, while genetic disorders were viewed as representing a small minority of the total clinical load, it is now becoming clear that the opposite is true: few clinical conditions are totally without some genetic influence.

The physician will soon need to be as familiar with genetic testing as he/she is with routine hematology and biochemistry analysis. While rapid and routine testing in molecular genetics is still an evolving field, in many situations such tests are already routine and represent essential adjuncts to clinical diagnosis (a good example is cystic fibrosis).

This series of monographs is intended to bring specialists up to date in molecular genetics, both generally and also in very specific ways that are relevant to the given specialty. The aims are generally two-fold:

(i)   to set the relevant specialty in the context of the new genetics in general and more specifically

(ii)   to allow the specialist, with little experience of genetics or its nomenclature, an entry into the world of genetic testing as it pertains to his/her specialty

These monographs are not intended as comprehensive accounts of each specialty — such reference texts are already available. Emphasis has been placed on those disorders with a strong genetic etiology and, in particular, those for which diagnostic testing is available.

The glossary is designed as a general introduction to molecular genetics and its language.

The revolution in genetics has been paralleled in recent years by the information revolution. The two complement each other, and the World Wide Web is a rich source of information about genetics. The following sites are highly recommended as sources of information:

1. PubMed. Free on-line database of medical literature.
    http://www.ncbi.nlm.nih.gov/PubMed/

2. NCBI. Main entry to genome databases and other information about the human genome project.
    http://www.ncbi.nlm.nih.gov/

3. OMIM. On line inheritance in Man. The On-line version of McKusick's catalogue of Mendelian Disorders. Excellent links to PubMed and other databases.
    http://www.ncbi.nlm.nih.gov/omim/

4. Mutation database, Cardiff.
    http://www.uwcm.ac.uk/uwcm/mg/hgmd0.html

5. National Coalition for Health Professional Education in Genetics. An organization designed to prepare health professionals for the genomics revolution.
    http://www.nchpeg.org/

6. Finally, a series of articles from the *New England Journal of Medicine*, entitled Genomic Medicine, has been made available free of charge on their website.
    http://www.nejm.org/

Eli Hatchwell
Cold Spring Harbor Laboratory

Preface

The most prevalent genetic diseases in the field of otorhinolaryngology are associated with sensorineural hearing impairment (SNHI) or concern nonsyndromic types of inner-ear hearing impairment. Indeed, the prevalence of SNHI competes with that of the most common chronic diseases.

The first insight that hearing impairment could be inherited appeared in the second edition of Politzer's *Lehrbuch der Ohrenheilkunde*, published in 1887. Direct or dominant inheritance was distinguished from indirect or recessive inheritance. Politzer based his conclusions on the work of another German author, Hartmann (1880). Prior to this, Wilde (1853), Liebreich (1861), and Uchermann (1869) had all found evidence that hearing impairment could be hereditary. In the same century, syndromes in which hearing impairment was a main feature were first described. Examples included Usher syndrome, branchio-oto-renal syndrome, Pendred syndrome, Treacher Collins' syndrome, and osteogenesis imperfecta tarda.

Although the majority of hereditary hearing impairments are nonsyndromic, most early attention was paid to syndromic hearing disorders because these can be differentiated clinically on the basis of their associated characteristics. With the introduction of the audiometer in the late 1930s, however, it became easier to describe and differentiate nonsyndromic forms of hearing impairment. Nevertheless, descriptions of families with nonsyndromic autosomal dominant hearing impairment were not published until the 1950s. Similarly, congenital forms of hearing impairment have received more attention than late-onset hearing impairments. The textbooks *The Causes of Profound Childhood Deafness* by George Fraser (1976) and *Genetic and Metabolic Deafness* by Konigsmark and Gorlin (first published in 1976) illustrate the steady increase in knowledge regarding hereditary syndromes with hearing impairment.

Genetics for ENT Specialists focuses on otorhinolaryngologic diseases for which the causative gene has been identified, and highlights the relevance of syndromic diseases to otorhinolaryngologic practice. The five chapters give complete descriptions of syndromes that extend beyond the field of otorhinolaryngology – in the interest of good clinical practice, most diseases require a multidisciplinary approach.

In addition to its role as a useful summary of selected otorhinolaryngologic diseases, we have enhanced the use of this book in daily practice by including "diagnostic" tables in the introduction to the inherited SNHI section; in nonsyndromic SNHI, the inheritance pattern, audiogram shape, onset age, and progression of hearing impairment all play a major role. Descriptions of the anatomical site or organ system that may be involved in each syndrome are also provided in table form.

The field of genetics has expanded rapidly during recent decades, and particularly over the last 10 years. This comprehensive text will keep the reader abreast of these new developments when identifying and treating diseases in daily practice. In addition, the addresses of several excellent websites are provided in the general introduction to ensure that the reader can access the most recent progress in various diseases.

Dirk Kunst, Hannie Kremer, and Cor Cremers

Acknowledgments

A special word of thanks goes to Ronald Admiraal for writing the sections on CHARGE association and the 22Q11 deletions, Emmanuel Mylanus for writing the section on cochlear implantation, Ronald Pennings for writing the sections on DFNA9 and Wolfram syndrome, Jeroen Jansen for contributing to the chapter dealing with paragangliomas, Robbert Ensink for contributing to the section on OXPHOS deficiencies, and Martijn Kemperman for contributing to the section on DFNB1. Dirk Kunst is the main author of the remaining sections. Hannie Kremer is responsible for providing the molecular genetic information.

We thank Patrick Huygen for reviewing the section entitled "Inherited nonsyndromic SNHI" and for providing most of the figures for this section. We also thank Philip van Damme and Ronald Pennings for reviewing the sections on craniosynostoses and Usher syndrome, respectively, and Greet Vantrappen for discussions regarding the section on 22Q11 deletions and DiGeorge syndrome in particular

Contents

Inherited nonsyndromic hearing impairment:
X-linked inheritance

2. Inherited syndromic hearing impairment

1. Inherited Diseases in Otology

Inherited nonsyndromic hearing impairment: recessive inheritance

Inherited nonsyndromic hearing impairment: X-linked inheritance

Genetics for ENT Specialists

Introduction: Inherited Sensorineural Hearing Impairment

The identification of genes that contribute to hearing and balance is helping to elucidate the molecular biology of the inner ear. In time, this research will lead to treatment strategies that prevent or stop the progression of hearing impairment – for example, by gene therapy. For this reason, it is important to keep abreast of new developments in the field of molecular biology of the inner ear.

Nowadays, approximately 1 in 1,000 neonates is severely hearing impaired, ie, with bilateral hearing thresholds of ≥80 dB. In at least half of these cases, the cause is inherited. The mode of inheritance can be autosomal recessive (70%–80% of patients), autosomal dominant (20%–30%), or X-linked (1%–2%); mitochondrial inherited sensorineural hearing impairment (SNHI) has also recently been described. In approximately 70% of hereditary cases, no other stigmata related to SNHI can be recognized – these types of hearing impairment are classified as nonsyndromic.

The above-mentioned data are mostly related to profound early childhood hearing impairment (prelingual phase). In the majority of patients with autosomal dominantly inherited hearing impairment, however, the age of onset is usually after early childhood (postlingual phase). The prevalence of postlingual SNHI in western Europe – with an average hearing threshold of >25 dB – is approximately 1% in young adults, about 10% up to the age of 60 years, and almost 50% at 80 years. The degree to which hereditary causes contribute to hereditary postlingual hearing impairment, and the prevalence of the different modes of inheritance, are unknown. Age-related hearing impairment is considered to be multifactorial, and is the result of both genetic and environmental factors.

Since 1992, the study of the molecular genetics of hereditary SNHI has made considerable progress due to improved and high-throughput methods, the human genome project, and studies in mouse models. It is becoming possible to make genetic diagnoses based on molecular tests in an increasing number of otologic disorders. This is of considerable importance, because different genotypes might lie behind (strikingly) similar phenotypes. This has been illustrated in autosomal dominant nonsyndromic SNHI, for which, until recently, only eight types could be distinguished clinically (according to Golin et al. [1995]). Nowadays,

DFNA no.	Gene	No. of families	Age of onset	Type of SNHI for the younger age group	Severity of impairment and progression[a]	Vestibular involvement	Remarks	Page no.
DFNA1	DIAPH1	1	5–20 years	Low frequencies	All frequencies Progressive to profound SNHI	Not described	May be Ménière-like	14
DFNA2	KCNQ4	>10	Early childhood	Downward-sloping audiogram	All frequencies Progressive[b]	Hyperreactivity in some	–	16
	GJB3 (CX31)	2	>20 years	High frequencies	High frequencies Progressive, causing mild to moderate SNHI	Not described	–	18
DFNA3	GJB2 (CX26), W44C mutation	2	Prelingual	Downward-sloping audiogram	High frequencies may show profound SNHI Mostly stable	Not described	–	19
	GJB2 (CX26), C202F mutation	1	Late childhood (postlingual)	Predominantly high frequencies	Progressive to mild to moderate SNHI	Not described	–	20
	GJB6 (CX30)	1	Unknown	Mid and high frequencies	Variable	Not described	–	21
DFNA5	DFNA5	2	5–15 years	High frequencies	Progressive – finally, all frequencies are moderately to profoundly affected	No vestibular test abnormalities	–	21
DFNA6/14/38	WFS1	5–10	Congenital	Low frequencies	Mostly stable, moderate SNHI	Vestibular function generally intact	–	23
DFNA8/12	TECTA	<10	Prelingual or early childhood	U-shaped audiogram, moderate SNHI	Usually stable	Not described	–	25
	TECTA, C1619S mutation	1	Prelingual	Predominantly high frequencies, mild SNHI	Progressive from mild to moderate SNHI	Symptoms reported	–	25

DFNA no.	Gene	No. of families	Age of onset	Type of SNHI for the younger age group	Severity of impairment and progression[a]	Vestibular involvement	Remarks	Page no.
DFNA9	COCH	15	Mid life	High frequencies	All frequencies Progressive to profound SNHI	Hyporeflexia and finally areflexia	May be Ménière-like	27
DFNA10	EYA4	3	0–40 years	Mild to moderate, may affect various frequencies	All frequencies None or some progression, causing moderate to severe SNHI	Not described	–	29
DFNA11	MYO7A	1	0–20 years	Flat or downward-sloping audiogram, mild to moderate SNHI	Minor progression, causing moderate SNHI	Decreased caloric tests in the elderly, without symptoms	–	31
DFNA13	COL11A2	2	0–30 years	U-shaped audiogram, mild to moderate SNHI	Usually stable	Caloric abnormalities, no symptoms	–	32
DFNA15	POU4F3	1	10–30 years	High frequencies Progressive to moderate	Progressive to moderate to severe SNHI	No symptoms	–	34
DFNA17	MYH9	1	>10 years	Mild, high-frequency SNHI	All frequencies Progressive, causing moderate to severe SNHI	No symptoms	–	35
DFNA22	MYO6	1	Childhood, postlingual	–	All frequencies Progressive, causing moderate to profound SNHI	No symptoms	–	36
DFNA36	TMC1	1	5–10 years	–	All frequencies Highly progressive to profound SNHI	No symptoms	–	37
DFNA39	DSPP	2	20–30 years	Mild, high-frequency SNHI	High frequencies Progressive, causing moderate to severe SNHI	Symptoms reported	–	38

Table 1. Summary of the characteristics of hearing impairment associated with mutations in the various DFNA loci. Numbers in the last column refer to the page with a full description of the corresponding locus. This table may be used for diagnostic purposes. [a]Progression beyond the P50 of presbyacusis. [b]Belgian family shows progression only at the high frequencies. SNHI: sensorineural hearing impairment.

DFNB no.	Gene	No. of families	Age of onset	Type of SNHI for the younger age group	Severity of impairment and progression[a]	Vestibular involvement	Remarks	Page no.
DFNB1	GJB2 (CX26)	High prevalence	Prelingual	**Ranging from mild to profound**	Progression in a minority of patients	None in tested patients	–	42
DFNB2	MYO7A	3	**Congenital to 16 years**	Severe to profound SNHI	Not applicable	Reduced or absent vestibular function	–	43
DFNB4/ EVA	SLC26A4	>10	Congenital	Moderate to severe **fluctuating** SNHI	Often progressive to profound SNHI	Vestibular dysfunction is common	**May be Ménière-like**	45
DFNB8/10	TMPRSS3	Frequently found	Congenital	Severe to profound SNHI	Not applicable	No symptoms	–	46
	TMPRSS3	1	0–12 years	–	**Extremely progressive, causing profound SNHI within 4–5 years**	Not described	–	46

Table 2. Most DFNB loci are related to hearing impairment that has a congenital or prelingual onset and is severe and mostly profound. DFNB loci with more exceptional characteristics (printed in **bold**) are listed in the table. Numbers in the last column refer to the page with a full description of the corresponding locus. [a]Progression beyond the P50 of presbyacusis. SNHI: sensorineural hearing impairment.

approximately 40 different loci are known to be responsible for this disorder.

The locus on the chromosome that harbors a gene involved in nonsyndromic autosomal dominant hearing impairment is specified by the prefix "DFNA" (see **Table 1**). Nonsyndromic autosomal recessive hearing impairment carries the prefix "DFNB" (see **Table 2**), while X-linked forms of nonsyndromic hearing impairment are prefixed by "DFN". About 35 DFNB and five DFN loci are known. In addition, three mitochondrial genes involved in nonsyndromic mitochondrial hereditary hearing impairment have been described. Many more loci are expected to be identified in the future. Currently, approximately 80 loci causing nonsyndromic SNHI are known; however, only around 28 genes have been identified. Altogether, over 130 genes have been identified for more than 400 genetic syndromes involving hearing impairment. Much of these data are available via the Hereditary Hearing Loss Homepage: http://www.uia.ac.be/dnalab/hhh/

Cochlear implantation

Cochlear implantation (CI) is a method of auditory rehabilitation for profoundly hearing-impaired patients for whom conventional amplification is insufficient. Multichannel CI is commonly performed in both adults and children. After a routine mastoidectomy and posterior tympanotomy, the cochlea is opened near the round window and an electrode array is inserted into the scala tympani. An external speech processor converts speech into an electric current via a specific algorithm; the electric current is subsequently transmitted to an internal receiver connected to the intracochlear electrode array. Thus, the cochlear nerve fibers are directly stimulated.

In the early days of CI, the majority of patients who benefited from this procedure suffered from postmeningitic profound bilateral SNHI. Nowadays, an increasing number of children who are scheduled for CI have congenital profound hearing impairment, and about half of these children have an autosomal recessive or autosomal dominant type of inherited deafness. Amongst these patients, many have profound hearing impairment associated with a specific genetic syndrome.

Genetic syndromes

Schuknecht (1980) noted that genetic and viral etiologies may be distinguished histologically. With genetic defects, the inner-ear structures were found to be poorly formed, but the cochlear nerve fibers were intact;

first trimester infections, however, produced atrophy of previously normal structures, including the nerve fibers. Malformations of the inner-ear structures may occur in both genetic syndromes and in isolated nonsyndromic hearing impairment. Twenty percent of all cases of congenital profound hearing impairment have bony abnormalities of the labyrinth.

In order to classify the various malformations and correlate surgical issues and rehabilitation outcome with certain types of malformation, most reports make use of the classification based on embryonic development suggested by Jackler et al (1987). During growth of the embryo, the developmental stage at which the cochlea is arrested correlates with the degree of severity of the malformation. Thus, a malformation of the cochlea may vary from total aplasia to severe cochlear hypoplasia, mild cochlear hypoplasia (basal turn only), common cavity, severe incomplete partition, mild incomplete partition, or a subnormal small cochlea that may or may not reach a full 2.5 turn. Various membranous abnormalities can also be detected by imaging. Cochlear malformation may present with a variety of bony abnormalities of the vestibule or semicircular canals, or an enlarged vestibular aqueduct, which is the most common inner-ear malformation (Mafee, 1992).

Genetic defects can produce a wide variety of malformations. Specific genes may control the development of separate inner-ear structures; on the other hand, a single gene defect can produce a wide variety of malformations, depending on the time at which these developmental effects occur (see **Table 3**) (Smith, 1998). Examples of this include the following syndromes (see **Table 4**).

- Waardenburg syndrome comprises an etiologically heterogeneous group. Features of this syndrome include pigmentary anomalies, such as a white forelock and heterochromia irides. Morphologic malformations of the labyrinth and membranous defects of the cochlea and vestibular system have also been reported. Aplasia of the semicircular canals is the most common characteristic, and frequently involves the posterior canal.

- Pendred syndrome is characterized by hearing impairment and an organification defect of iodine that leads to thyroid enlargement and goiter. Pendred syndrome and enlarged vestibular aqueduct syndrome

Abnormal area	Syndromes with a high prevalence of the feature	Syndromes with a low prevalence of the feature
Ear (external) and region	BOR, CHARGE, Noonan, oculo-auriculo-vertebral spectrum, Saethre–Chotzen, Treacher Collins	Apert, Crouzon, Pfeiffer
Eye (region)	Apert, CHARGE, Crouzon, Mohr Tranebjaerg, neurofibromatosis type II, Noonan, Norrie, osteogenesis imperfecta, osteopetrosis, OXPHOS, Pfeiffer, Saethre–Chotzen, Stickler, Treacher Collins, Usher, Waardenburg, Wolfram	Alport, BOR, oculo-auriculo-vertebral spectrum
Mouth and region, including teeth (soft tissue)	Crouzon, Noonan, oculo-auriculo-vertebral spectrum, osteogenesis imperfecta, osteopetrosis, Pfeiffer, Saethre–Chotzen, Stickler	Apert, CHARGE, Noonan, Treacher Collins, Waardenburg
Nose and associated area	Apert, CHARGE, Crouzon, Noonan, Pfeiffer, Saethre–Chotzen, Stickler, Waardenburg	–
Throat and associated region	–	CHARGE, Pfeiffer
Neck and associated region	BOR, Noonan	–
Heart	CHARGE, Jervell and Lange-Nielsen, Noonan, OXPHOS	Apert, oculo-auriculo-vertebral spectrum, osteogenesis imperfecta, Saethre–Chotzen, Stickler, Treacher Collins
Hands/feet	Apert, Pfeiffer, Saethre–Chotzen	Osteogenesis imperfecta
Genitourinary	Alport, BOR, CHARGE, Noonan, osteopetrosis with renal tubular acidosis, Wolfram	Apert, Norrie, oculo-auriculo-vertebral spectrum, Saethre–Chotzen, Waardenburg
Skin	Apert, neurofibromatosis type II, Waardenburg	–
Thyroid gland	Pendred	–
Nervous system, including retardation	Apert, CHARGE, Mohr Tranebjaerg, neurofibromatosis type II, OXPHOS, Wolfram	Crouzon, Pfeiffer, Noonan, Norrie, oculo-auriculo-vertebral spectrum, osteopetrosis, Saethre–Chotzen, Waardenburg
Skeleton/joints (neuro- and viscero-cranium)	Apert, Crouzon, oculo-auriculo-vertebral spectrum, osteogenesis imperfecta, osteopetrosis, Pfeiffer, Saethre–Chotzen, Stickler, Treacher Collins, Waardenburg type III	–
Skeleton/joints (other)	Apert, Noonan, osteogenesis imperfecta, osteopetrosis, Stickler, Waardenburg type III	Crouzon, CHARGE, oculo-auriculo-vertebral spectrum, Pfeiffer
(Additional) conductive hearing impairment, OME excluded	BOR, Crouzon, oculo-auriculo-vertebral spectrum, osteogenesis imperfecta, Treacher Collins	Apert, Noonan, osteopetrosis, Pfeiffer, Saethre–Chotzen

Table 3. List of the various anatomical sites or organ systems that can be affected in syndromes associated with hearing impairment. A distinction is made between high and low prevalence of the additional symptoms. The last row indicates syndromes that might exhibit an (additional) conductive hearing impairment. Only syndromes described in this book are included in the table. BOR: branchio-oto-renal syndrome; CHARGE: coloboma of the eye (C), heart anomaly (H), atresia choanae (A), retardation of growth and/or development (R), genitourinary anomalies (G), ear anomaly, hearing impairment, or deafness (E); OME: otitis media with effusion; OXPHOS: oxidative phosphorylation deficiencies.

Syndrome	Gene	Chromosome	Hearing impairment	Vestibular dysfunction	Cochlear malformation	Vestibular malformation	Comment
Waardenburg type I	PAX3	2q35	Sensorineural	Yes	Membranous	Aplasia of posterior SCC	–
Pendred	SLC26A4	7q21–q34	Sensorineural	Yes	Incomplete partition	EVA	–
Wildervanck	Unknown	Unknown	Mostly sensorineural	Unknown	Yes	SCC may be absent	IAC may be abnormal
Branchio-oto-renal	EYA1	8q13.3	Sensorineural: mixed or conductive	No	Reduced size	Hypoplasia, EVA	Ossicular chain malformation, oval window aplasia; possible progressive hearing impairment
DiGeorge	Unknown	22q11.2	Sensorineural	Unknown	Incomplete partition	–	External and middle-ear malformations may be present

Table 4. Inner-ear malformations that have been identified in specific genetic syndromes associated with profound hearing impairment. EVA: enlarged vestibular aqueduct; IAC: internal auditory canal; SCC: semicircular canal.

share the same causative gene, and the mutations that lead to these two diseases occur in the same location on the gene; thus, both syndromes are part of a spectrum. Inner-ear anomalies – such as hypoplasia of the cochlea and an enlarged vestibular aqueduct – are common. An enlarged vestibular aqueduct seems to be ubiquitous and in 80% of cases it is associated with Pendred or enlarged vestibular aqueduct syndromes. The hearing impairment is sensorineural and can be congenital or prelingual; there may also be a sudden hearing impairment that progresses over time. Incomplete partition of the cochlea and enlarged vestibular aqueducts also occur among inner-ear malformations.

- In some genetic syndromes, external and middle-ear anomalies may be present in combination with anomalies of the inner ear; however this mixed dysplasia is rare. Wildervanck's syndrome comprises the Klippel–Feil anomaly (fusion of the cervical vertebrae), Duane retraction (recession of the eye when gazing laterally), and SNHI. The syndrome is a result of a heterogeneous, multifactorial trait. A variety of inner-ear malformations have been identified, including abnormalities of the cochlea, vestibule, semicircular canals, and the internal auditory canal.

- Branchio-oto-renal syndrome is characterized by branchial pits, cysts, abnormal shape of the pinna, preauricular pits, renal malformation (including small or missing kidneys), and hearing impairment. Ossicular chain malformations and oval window aplasia are also sometimes present. Inner-ear findings include a small cochlea or more extensive cochlear hypoplasia. Hypoplasia of the semicircular canals and an enlarged vestibular aqueduct are common.

- DiGeorge syndrome includes immune deficiency, hypocalcemia, heart defects, and distinctive facial features. External and middle-ear malformations may be present in combination with incomplete partition of the cochlea.

Surgical aspects
When assessing the technical feasibility of CI in profoundly deaf patients with a malformed cochlea, it is of foremost importance to determine preoperatively whether there is sufficient cochlear lumen for electrode placement and to rule out eighth nerve aplasia or hypoplasia. Therefore, computed tomography (CT) and magnetic resonance imaging (MRI) are

of great importance in this particular group of implant candidates. Aplasia of the eighth nerve needs to be ruled out with MRI, especially in patients with a narrow internal auditory canal that has been visualized on a CT scan or in patients with CHARGE association. CT scanning alone is not sufficient for this, and submillimetric MRI scanning in the axial and parasagittal planes is necessary.

CI in patients with genetic syndromes who have normal or near-normal inner-ear anatomy is straightforward. As a consequence, CI does not pose a surgical problem in – for instance – patients with Usher syndrome type I with retinitis pigmentosa and congenital deafness with vestibular areflexia, both of which are due to hair cell dysfunction.

In patients with inner-ear malformations, certain aspects of the surgical procedure are of importance. Although surgery is considered feasible, CI might be more complicated in these patients. This is a result of the abnormal anatomy of the temporal bone, including a possible aberrant course of the facial nerve and the occurrence of cerebrospinal fluid (CSF) gusher. Instances of facial nerves following an aberrant path have been reported in 16% of inner-ear malformations in general, and noted more frequently in patients with severe malformations, such as a common cavity or severe hypoplastic cochleas (Hoffman, 1997). The incidence of CSF gushers is 40% – irrespective of the severity of the inner-ear malformation – and most cases can be easily treated by packing the cochleostomy thoroughly with soft tissue so that lumbar drainage of CSF is rarely necessary. Postoperative CSF leakage poses a risk for meningitis, which may even occur several months postoperatively. A CSF gusher may be predicted preoperatively by CT and MRI, ruling out a patulous cribriform area between the cochlea and internal auditory canal.

Treatment outcome
The results of CI may be excellent in patients with a genetic syndrome associated with profound hearing impairment. For instance, in patients with Usher syndrome type I, the results are comparable with those obtained in other prelingual deaf patients (Hinderink, 1994). As with all CIs, the outcome for the patient depends on confounding factors, such as age at implantation and duration of deafness.

If an inner-ear malformation is detected, CI is often still feasible. Several clinics have reported the benefit of CI in these children. This is certainly true for children with labyrinthine abnormalities and normal cochleae, and for patients with an enlarged vestibular aqueduct. Generally, in patients with mild cochlear deformities – such as mild or severe incomplete partition – full insertion of the electrode array is possible, and results comparable with those obtained in profoundly deaf patients with normal cochleae can be obtained.

Inherited Nonsyndromic Hearing Impairment: Dominant Inheritance

DFNA1

MIM   124900

Clinical features   Progressive, low-frequency SNHI, which finally affects all frequencies. No detailed phenotypic study has been reported for the one DFNA1 family so far described; audiograms of only six patients have been published (see **Figure 1**). Because only one DFNA1 family has been described to date, nothing is known about interfamilial variation. SNHI presents after childhood, predominantly in the low frequencies (250 Hz, 500 Hz, 1,000 Hz). Impairment progresses through adolescence and losses of up to 70 dB may be encountered at all frequencies. Adults often have hearing impairment in excess of 100 dB at all frequencies. This SNHI first affects the low frequencies and may present with tinnitus at onset.

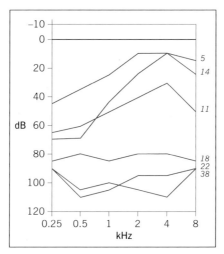

Figure 1. Audiograms of six affected individuals in the DFNA1 family. Numbers in italics refer to ages (years).

Age of onset	Clinical symptoms have been reported at as early as 5 years and up to about 20 years. However, subclinical SNHI at low frequencies may be present at birth.
Epidemiology	Only one family with this condition is known.
Inheritance	Autosomal dominant
Chromosomal location	5q31
Genes	*DIAPH1* (diaphanous)
Mutational spectrum	Only one mutation has been identified – a nucleotide substitution in a splice site leading to a frame shift and protein truncation.
Effect of mutation	The diaphanous protein is likely to be involved in regulating the polymerization of actin in hair cells and pilar cells in the cochlea. Actin is the major component of the cytoskeleton and of the stereocilia of hair cells. The mutation might lead to reduced amounts of protein. Alternatively, aberrant protein may cause hearing impairment by disturbing the above-mentioned regulation of actin polymerization.
Diagnosis	Patients with DFNA1 may have been previously diagnosed with a Ménière-like disease. Low-frequency SNHI in a younger individual with older relatives showing severe progression suggests DFNA1 involvement. Mutation analysis may confirm the diagnosis.
Counseling issues	Patients show normal development. Penetrance of the disease is complete in the one, very large DFNA1 family so far described. The pedigree of the family dates back to about 1740 and comprises nine generations. Two hundred family members were evaluated audiometrically.

DFNA2

MIM	600101
	This locus harbors at least two genes for hearing impairment: *KCNQ4* and *GJB3*.

(a) DFNA2, *KCNQ4* Gene

MIM	603537
Clinical features	Symmetric, progressive, high-frequency SNHI. At onset, most families also show low-frequency SNHI to a lesser extent, depending on the type of mutation. Progression is usually found at all frequencies, leading to the involvement of all frequencies at a more advanced age. Several phenotypes are known and these depend on the location of the mutation. Most mutations (dominant-negative effect) lead to progressive SNHI affecting the high frequencies and, to a lesser degree, the low frequencies. Progression in these families is about 1 dB/year at all frequencies (see **Figure 2**). One mutation, Q71fs, found in a Belgian family (haploinsufficiency), led to progressive deterioration exclusively at the high frequencies.
Age of onset	Possibly congenital for the high frequencies in most families, usually resulting in subclinical SNHI in the prelingual phase. As a result of this, postlingual SNHI can be detected clinically.
Epidemiology	More than 10 unrelated families have been reported (Dutch, Belgian, Japanese, French, and American). DFNA2 is the dominantly inherited locus, with one of the highest numbers of families linked to it.
Inheritance	Autosomal dominant
Chromosomal location	1p34
Genes	*KCNQ4* (potassium channel, voltage-gated, KQT-like subfamily, member 4)
Mutational spectrum	Missense mutations and a small deletion have been found. The W276S mutation, which has been found in three unrelated families, seems to represent a mutational hot spot.

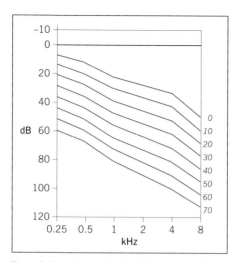

Figure 2. Typical age-related audiograms of six families with mutations confined to the pore region of *KCNQ4*. Numbers in italics refer to ages (years). Figure courtesy of P Huygen.

Effect of mutation

Most of the mutations have a dominant-negative effect, leading to disturbance of the homo- and possibly heterotetramers of the *KCNQ4* subunits. The truncating mutations might lead to haploinsufficiency. *KCNQ4* forms a K^+ channel in the outer hair cells, which is responsible for the removal of K^+ from these cells after depolarization by K^+ influx via the apical membrane. A high K^+ concentration in the hair cells may finally lead to their degeneration. Various studies have shown expression of *KCNQ4* not only in the hair cells, but also in many nuclei of the auditory pathway. Thus, DFNA2 might also include a central defect.

Diagnosis

A *KCNQ4* mutation should be suspected in any patient with dominantly inherited, high-frequency SNHI. Mutation analysis may confirm the diagnosis.

Counseling issues

Because of considerable inter- and even intrafamilial variability regarding onset, progression, and severity, individual phenotypic counseling must be handled with care. In most individuals, speech and language acquisition are not jeopardized. At a given SNHI, speech recognition scores in these patients seem to be significantly better than in patients with age-related hearing impairment (presbyacusis) or in DFNA9 patients (see p. 27). The penetrance of DFNA2 due to mutations in *KCNQ4* is complete.

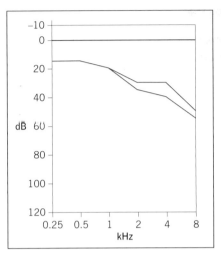

Figure 3. Audiogram of both ears of a 56-year-old man suffering from sensorineural hearing impairment caused by a mutation in *GJB3*.

(b) DFNA2, *GJB3* Gene

MIM	603324
Clinical features	Progressive, high-frequency SNHI accompanied by tinnitus. Progression may cause mild to moderate SNHI (see **Figure 3**).
Age of onset	From about 20 years onwards
Epidemiology	This syndrome is known in two Chinese families.
Inheritance	Autosomal dominant
Chromosomal location	1p35.1
Genes	*GJB3* (gap-junction protein β-3; also known as *CX31*[connexin 31])
Mutational spectrum	Nucleotide substitutions leading to amino-acid changes have been reported.
Effect of mutation	This mutation leads to a disturbance of intercellular metabolic and electrical communication, resulting in hearing impairment.

Diagnosis	Mutation analysis may confirm the diagnosis.
Counseling issues	Penetrance may be reduced. This gene has also been described as being involved in recessively inherited deafness in three patients of two nonconsanguineous Chinese families. An early-onset bilateral hearing impairment was found, varying from moderate to severe, and the audiogram was flat. Vestibular function tests were normal.

DFNA3

MIM	601544, with the following three subtypes:

(a) DFNA3, W44C Mutation

MIM	121011
Clinical features	Symmetric, moderate to severe SNHI, predominantly affecting the high frequencies (70–120 dB) although low frequencies (15–90 dB) may also be involved. About 50% of patients show no hearing ability at 120 dB. SNHI is only slightly progressive in most patients, and nonprogressive in some.
Age of onset	Prelingual
Epidemiology	This syndrome is known in two French families.
Inheritance	Autosomal dominant
Chromosomal location	13q11–q12
Genes	GJB2 (gap-junction protein β-2; also known as CX26 [connexin 26])
Mutational spectrum	The nucleotide substitution 132G>C leads to the amino-acid substitution W44C.
Effect of mutation	There is a dominant-negative effect on the function of the gap junction, which is involved in K^+ recycling. Mutations lead to disturbances of intercellular metabolic and electrical communication. GJB2 expression

has been demonstrated in the stria vascularis, basilar membrane, limbus, and spiral prominence of the cochlea.

Diagnosis	Mutation analysis will lead to the diagnosis.
Counseling issues	Intrafamilial variability is low. Certain mutations may result in SNHI in combination with palmoplantar keratoderma, making the disease syndromic. Mutations in the *GJB2* gene also cause DFNB1 (see p. 42).

(b) DFNA3, C202F Mutation

MIM	121011
Clinical features	SNHI occurs predominantly at the high frequencies in the first decade, but progresses to the mid frequencies at the age of 10–50 years. Finally, mild to moderate SNHI occurs.
Age of onset	Late childhood (postlingual), mostly between 10 and 20 years
Epidemiology	This syndrome is known in only one French family.
Inheritance	Autosomal dominant
Chromosomal location	13q11–q12
Genes	*GJB2* (gap-junction protein β-2; also known as CX26 [connexin 26])
Mutational spectrum	C202F mutation (the result of a 605G>T mutation)
Effect of mutation	There is a dominant-negative effect on the function of the gap junction, which is involved in K^+ recycling. Disturbances of intercellular metabolic and electrical communication occur in this condition. *GJB2* expression has been demonstrated in the stria vascularis, basilar membrane, limbus, and spiral prominence of the cochlea.
Diagnosis	Mutation analysis will lead to the diagnosis.
Counseling issues	Intrafamilial variability has been found. Certain mutations may result in SNHI in combination with palmoplantar keratoderma, making the disease syndromic. Mutations in *GJB2* also cause DFNB1 (see p. 42).

(c) DFNA3, *GJB6* Gene

MIM	604418
Clinical features	Bilateral mid- and high-frequency SNHI. The affected mother of two affected children showed 20–50 dB SNHI. Her son suffered from progressive SNHI above 500 Hz and her daughter had profound SNHI.
Age of onset	Unknown
Epidemiology	This syndrome is known in only one Italian family.
Inheritance	Autosomal dominant
Chromosomal location	13q12
Genes	*GJB6* (gap-junction protein β-6; also known as *CX30* [connexin 30])
Mutational spectrum	There is a nucleotide substitution leading to the amino-acid change T5M.
Effect of mutation	This mutation has a dominant-negative effect on the function of the ion channel. Disturbances of intercellular metabolic and electrical communication are characteristic. *GJB6* expression has been demonstrated in the lateral wall of the cochlea.
Diagnosis	Mutation analysis will lead to the diagnosis.
Counseling issues	Expression is variable. Certain mutations may be found in hidrotic ectodermal dysplasia; these are sporadically accompanied by SNHI, in which case the disease is syndromic.

DFNA5

MIM	600994
Clinical features	Progressive, symmetric, high-frequency SNHI with low-frequency involvement at a later stage (see **Figure 4**). In the first phases of degeneration, impairment of the high tones proceeds from 0 to 60 dB (about 1–4 dB/year at 2–8 kHz), whereas low-tone hearing remains almost unaffected. When high-frequency impairment reaches 60 dB,

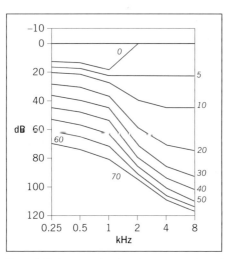

Figure 4. Typical age-related audiogram in a DFNA5 family, based on original data from EH Huizing (personal communication). Numbers in italics refer to ages (years). Figure courtesy of P Huygen.

low-frequency hearing also starts to deteriorate (around 1 dB/year at 0.25–1 kHz). By the end of the degenerative process, the high-tone impairment is maximal (>105 dB) while the low-tone impairment amounts to approximately 60 dB. Vestibular function and tests are normal.

Age of onset	Between 5 and 15 years of age
Epidemiology	This condition has been recognized in two large Dutch families. One of the two families contained more than 100 mutation carriers.
Inheritance	Autosomal dominant
Chromosomal location	7p15
Genes	*DFNA5* (deafness, autosomal dominant nonsyndromic sensorineural 5)
Mutational spectrum	A complex deletion/insertion mutation in intron 7 and a nucleotide substitution in this intron (IVS7-6C>G) have been reported. Both mutations lead to exon 8 being skipped in the messenger RNA and are predicted to lead to premature protein truncation.

Effect of mutation	*DFNA5* expression has been demonstrated in the cochlea; however, the physiologic function of the protein and consequently the direct effect of the mutation remain unknown. The protein may play a role in apoptosis.
Diagnosis	This condition is suggested by the presence of high-frequency SNHI with rapid deterioration of the lower thresholds at a later stage. Mutation analysis will confirm the diagnosis.
Counseling issues	Full penetrance has been reported. In these patients, speech recognition scores at a given SNHI (pure tone average [PTA] 1–4 kHz >70 dB) seem to be better than those among patients with age-related hearing impairment (presbyacusis). In the case of progression to profound deafness, cochlear implantation may be advised; this usually has a successful outcome.

DFNA6/DFNA14/DFNA38

(see also: Wolfram syndrome)

MIM	600965
Clinical features	In the first two families studied, nonprogressive, low-frequency SNHI was found, with losses of about 40–60 dB at 0.25–1 kHz in individuals aged <40 years (see **Figure 5**). Up to the age of approximately 40 years, SNHI appears to be stable. The only progression observed is caused by presbyacusis, finally leading to downsloping, nonprofound SNHI. Three other families showed similar SNHI in the younger age group. However, SNHI found at a more advanced age in the low frequencies (0.25–1 kHz) tended to progress beyond the P50 of presbyacusis. Minor inter- and intrafamilial variation in phenotypic expression was found among all the families studied.
Age of onset	This is probably a congenital condition.
Epidemiology	To date, mutations in the *WFS1* gene (see below) have been found to be the most common cause of inherited low-frequency hearing impairment. So far, more than 10 DFNA6 families and sporadic cases have been identified. The only other locus that causes low-frequency hearing impairment is DFNA1, which has been found in only one family.

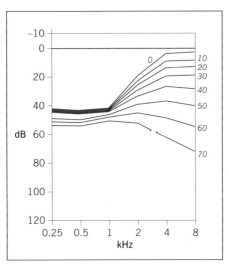

Figure 5. Mean age-related audiogram of two DFNA6/DFNA14 families. Numbers in italics refer to ages (years). Figure courtesy of P Huygen.

Inheritance	Autosomal dominant
Chromosomal location	4p16.1
Genes	*WFS1* (wolframin)
Mutational spectrum	Nucleotide substitutions leading to amino-acid substitutions and a small deletion causing the deletion of one amino acid have been described.
Effect of mutation	Mutations in *WFS1* are mainly noninactivating. However, the exact role and function of wolframin remain unknown. It probably plays a crucial role in the survival of specific endocrine and neuronal cells. Biochemical studies show that wolframin is an integral transmembrane protein, predominantly located in the endoplasmic reticulum.
Diagnosis	Dominantly inherited, low-frequency SNHI with losses of 40–60 dB at 0.25–1 kHz in a younger age group (<40 years) suggests the involvement of DFNA6/DFNA14/DFNA38. The phenotype of this syndrome can be differentiated from that of DFNA1 because the latter shows more progression and a more severe type of SNHI. Mutation analysis may confirm the diagnosis.

| Counseling issues | Full penetrance has been reported. Speech and language acquisition are not jeopardized in families with this condition. As there are only minor variations in this impairment, individual phenotypic counseling is feasible. The use of a hearing aid may not be necessary in younger individuals, because hearing at the higher frequencies is sufficiently preserved to allow speech recognition. |

DFNA8/DFNA12

DFNA8 and DFNA12 are the same disorder. It was initially thought that this form of autosomal dominant deafness involved two different loci. However, it is now understood that DFNA8 and DFNA12 concern the same locus and the same gene.

MIM	601543/601842
Clinical features	In a Belgian and an Austrian family with this condition, SNHI was nonprogressive (even below the P50 of presbyacusis) and moderate to moderately severe. Although all frequencies are affected, this disease affects the mid frequencies most severely, resulting in an average loss of approximately 50 dB at 0.5–2 kHz (see **Figure 6**). The hearing impairment tends be almost stable, even at a more advanced age (ie, when presbyacusis would be expected to have a major influence). Because of this only minor progression, SNHI will not become profound at a more advanced age. The phenotype of the Austrian family showed only minor interindividual variation, irrespective of age; however, the Belgian family showed considerable intrafamilial variability. CT and MRI scans of temporal bones did not show any gross structural abnormalities in the Belgian family. In a French family (mutation C1619S), prelingual, symmetric, and predominantly high-frequency SNHI was reported. The SNHI was progressive from mild to moderate, and the average rate of progression was approximately 0.7 dB/year at 0.5–4 kHz. Possible vestibular involvement was reported anamnestically.
Age of onset	Prelingual or during early childhood
Epidemiology	To date, less than 10 families with *TECTA* mutations (see below) have been identified.
Inheritance	Autosomal dominant

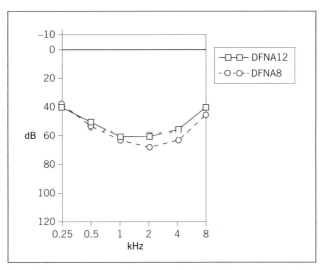

Figure 6. Mean audiograms of 17 affected individuals in a DFNA12 family and 10 affected individuals in a DFNA8 family. Figure courtesy of P Huygen.

Chromosomal location	11q22–q24
Genes	*TECTA* (α-tectorin)
Mutational spectrum	Nucleotide substitutions have been reported, all of which lead to the substitution of an amino acid, such as C1057S, C1619S, L1820F, G1824D, and Y1870C.
Effect of mutation	The protein α-tectorin is one of the major components of the tectorial membrane, which is responsible for the deflection of stereocilia in hair cells. This deflection depolarizes the hair cells, thus allowing the perception of sound. Mutations may result in the structure of the tectorial membrane being altered in a dominant-negative fashion, thereby causing SNHI.
Diagnosis	This condition is characterized by dominantly inherited, stable, mid-frequency SNHI, with an average loss of 50 dB at 0.5–2 kHz. The C1619S mutation gives a less specific audiogram. Mutation analysis may confirm the diagnosis.

Counseling issues	Speech development can be delayed in all families with this condition, especially when a hearing aid is not fitted at an early age. The penetrance of this disease is complete. Other mutations in the *TECTA* gene cause DFNB21 (see p. 49).

DFNA9

MIM	601369
Clinical features	Mid-life onset of progressive, high-frequency SNHI and vestibular impairment. At present, this is the only known type of DFNA to involve vestibular areflexia. Audiograms obtained in the first three to four decades of life show a flat to gently downward-sloping pure-tone threshold with virtually no impairment in the low-frequency range. After the fourth decade, progression affects all frequencies and the downward-sloping pure-tone threshold eventually progresses to profound deafness with some residual hearing at the lower frequencies (see **Figure 7**). Functional degeneration of the inner ear leads to impairment of the vestibulum, causing hyporeflexia, which eventually results in vestibular areflexia. The clinical symptoms of DFNA9 seem to mimic those of Ménière's disease, including clinical manifestations such as episodes of vertigo, hearing impairment, tinnitus, and aural fullness. These symptoms are probably related to asymmetry in vestibular dysfunction.
Age of onset	This condition is characterized by mid-life onset (approximately 35–40 years).
Epidemiology	Unknown, although about 15 families with this condition have been identified.
Inheritance	Autosomal dominant
Chromosomal location	14q12–q13
Gene	*COCH* (cochlin)
Mutational spectrum	Five missense mutations have been described for different families from The Netherlands, Belgium, the US, and Australia. All of these mutations

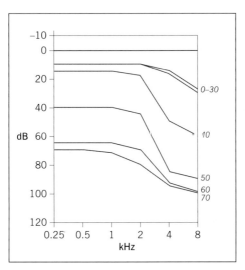

Figure 7. Typical age-related audiogram of a Dutch DFNA9 family. Numbers in italics refer to ages (years). Figure courtesy of P Huygen.

are located in a region containing four conserved cysteines near the amino terminus of *COCH*. The P51S mutation appears to be of major importance.

Effect of mutation How the mutations lead to severe inner-ear degeneration is unknown. In mice, the *coch* gene is expressed in the fibrocytes of the spiral limbus, the spiral ligament of the cochlea, and the fibrocytes of the connective tissue stroma of the crista ampullaris in the semicircular canals. The product of this gene is most likely an extracellular protein.

Diagnosis The clinical features and autosomal dominant inheritance pattern, which distinguishes DFNA9 from Ménière's disease, are the main features by which DFNA9 can be identified. Mutation analysis will confirm the clinical diagnosis in most patients.

Counseling issues Regular audiometric testing is advised from the age of 35 years. Where profound deafness has already developed, cochlear implantation is usually successful. The use of extra light in dark conditions will alleviate some of the difficulties associated with vestibular areflexia. In the south of The Netherlands and in Belgium, the P51S mutation seems to affect several families who all originate from the same ancestor (a founder effect).

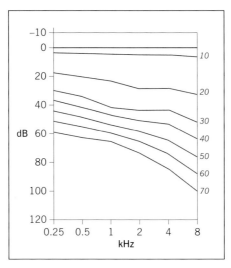

Figure 8. Mean age-related audiogram of two DFNA10 families. Numbers in italics refer to ages (years). Figure courtesy of P Huygen.

DFNA10

MIM	601316

Clinical features

One Belgian family with this condition showed progressive, symmetric SNHI. At onset, the mid frequencies are mainly affected or the audiogram might show a gentle downward slope. However, the impairment extends to all frequencies at a more advanced age. Initially, SNHI presents as a mild to moderate impairment that progresses to moderate or severe impairment after the age of about 55 years (see **Figure 8**). The average progression at 0.5–4 kHz is approximately 1.1 dB/year. There is considerable variability in the time of onset, severity, and shape of the audiogram. Tinnitus was reported in around 35% of cases in the Belgian family, which contained about 17 genetically affected individuals.

The phenotype of a second (American) family shows some similarities. At onset, SNHI is thought to present as a flat or gently sloping audiogram of about 50 dB, which is caused by early deterioration across all frequencies during the first decades of life. From the age of around 30 years onwards, progression is in line with presbyacusis and this finally results in a downward-sloping audiogram. An average progression of approximately 0.6 dB/year has been reported for all frequencies. At a given level of SNHI, speech recognition scores in these patients seem to be about the same as those in patients with presbyacusis.

No detailed data are available as yet for the third family (Norwegian) known to suffer from this condition. However, it has been suggested that the phenotypic details are much the same.

Age of onset
In the Belgian family, onset was postlingual, starting in the first to fourth decade (6–40 years, median 30 years). In the American family, postlingual onset was during the first three decades of life.

Epidemiology
This condition has been identified in one Belgian, one American, and one Norwegian family.

Inheritance
Autosomal dominant

Chromosomal location
6q22–q23.2

Genes
EYA4 (eyes absent 4)

Mutational spectrum
A small insertion and a nucleotide substitution have been described. Both are predicted to cause premature protein truncation.

Effect of mutation
Mutations in the *EYA4* gene are predicted to lead to either haploinsufficiency or a dominant-negative effect. *EYA4* encodes a transcriptional activator that interacts with other proteins to regulate early developmental events. Because of the progressive nature of this disease, *EYA4* probably also plays a survival role in the mature system, at least in the cochlea.

Diagnosis
This condition is characterized by postlingual, progressive (at least in the first decades), and mild to moderate (in younger individuals) SNHI. Mid-frequency SNHI may be found in the younger age group. Mutation analysis may confirm the diagnosis.

Counseling issues
There are no special recommendations.

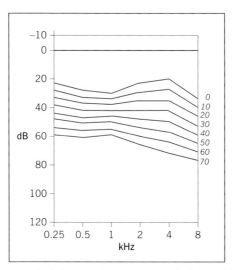

Figure 9. Typical age-related audiogram of a DFNA11 family.
Numbers in italics refer to ages (years). Figure courtesy of P Huygen.

DFNA11

MIM	601317
Clinical features	Symmetric, progressive SNHI involving all frequencies, with a gently sloping or flat audiogram shape (see **Figure 9**). The majority of those patients affected between the ages of 20 and 60 years have moderate SNHI. The mean annual threshold deterioration is calculated to be about 0.56 dB/year at all frequencies, with a trend to be less at the lower frequencies. Caloric vestibular tests may show decreased responses in the older age group (of five persons tested, three aged over 50 years had vestibular abnormalities); these patients, however, appear to be asymptomatic. No eye abnormalities or CT abnormalities of the temporal bone have been found.
Age of onset	Onset is postlingual, mainly in the second decade. However, subclinical onset in early childhood cannot be ruled out.
Epidemiology	This condition is known in one Japanese family.
Inheritance	Autosomal dominant

Chromosomal location	11q13.5
Genes	*MYO7A* (myosin VIIA)
Mutational spectrum	An in-frame 9-bp deletion of exon 22 has been found. Other mutations in this gene are responsible for nonsyndromic recessive SNHI (DFNB2; see p. 43) and Usher syndrome type IB (see p. 112).
Effect of mutation	Myosin VIIA is an actin-based motor protein, which is present in the cochlear hair cell stereocilia, the vestibular sensory cells, and the epithelium of the retinal pigment. Mutation of the gene affects the coiled-coil domain of myosin VIIA, which is important for dimerization of the protein. The mutation is likely to have a dominant-negative effect.
Diagnosis	Mutation analysis may reveal the diagnosis in patients with autosomal dominantly inherited, postlingual, moderate SNHI.
Counseling issues	Full penetrance has been reported. Individuals should be screened for pigmentary retinopathy.

DFNA13

MIM	601868
Clinical features	A Dutch family has shown combined mid- and high-frequency SNHI. Average thresholds of approximately 25 dB at 250, 500, and 4,000 Hz; 35–40 dB at 1,000, 2,000, and 6,000 Hz; and 50 dB at 8,000 Hz were found before the age from which presbyacusis begins to play a role. A typical feature in the mean audiogram of this family is the reset most often found at 4 kHz. Normal presbyacusis changes the typical shape of the audiogram into a downsloping type (see **Figure 10**). Caloric abnormalities were found in eight of 17 individuals; however, no obvious symptoms were apparent. An American family showed mid-frequency SNHI in the younger age group with thresholds of about 44 dB at 1, 2, and 4 kHz and approximately 29 dB at 0.25, 0.5, and 8 kHz.
Age of onset	Onset was between the ages of 0 and 30 years in the Dutch family. In the American family, it was thought to be prelingual.

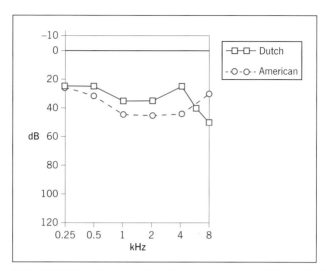

Figure 10. Mean audiogram of affected younger generation individuals of Dutch and American DFNA13 families. Figure courtesy of P Huygen.

Epidemiology	Two families with this condition are known (one American and one Dutch).
Inheritance	Autosomal dominant
Chromosomal location	6p21.3
Genes	*COL11A2* (collagen 11 α2-chain)
Mutational spectrum	Two nucleotide substitutions leading to missense mutations have been identified: R549C in the American family and G323E in the Dutch family.
Effect of mutation	Both mutations lead to an abnormal type XI collagen α2-chain, resulting in collagen fibril (type II collagen) disorganization in the tectorial membrane. This is predicted to interfere with sound conduction to the hair cells.
Diagnosis	This condition is characterized by mild to moderate SNHI that mainly involves the mid frequencies in young individuals (below 40–50 years) and an absence of progression. Mutation analysis may confirm the diagnosis.

| Counseling issues | Possible reduced penetrance has been reported in the Dutch family. Other mutations in *COL11A2* may cause Stickler syndrome type III, Weissenbacher–Zweymuller syndrome, and oto-spondylo-megaepiphysial dysplasia; affected persons should therefore be screened for features of these syndromes. |

DFNA15

MIM	602459
Clinical features	At onset, SNHI presents at the higher frequencies. Greater progression than would be expected on the basis of presbyacusis occurs at all frequencies, ie, 1.1 dB/year at 0.25–1 kHz and 2.1 dB/year at 2–8 kHz. At around the age of 50 years, moderate to severe SNHI occurs at all frequencies, giving a flat or downward-sloping audiogram (see **Figure 11**). There is intrafamilial variability regarding onset and progression. No vestibular symptoms have been reported.
Age of onset	The second or third decades of life
Epidemiology	This condition is known in one Israeli Jewish family with Italian roots.
Inheritance	Autosomal dominant
Chromosomal location	5q31
Genes	*POU4F3* (POU domain, class 4, transcription factor 3; also known as *BRN3C* [brain 3C])
Mutational spectrum	An 8-bp deletion (884del8) has been identified in the part of the gene encoding the POU homeodomain.
Effect of mutation	*POU4F3* is expressed in the inner ear, including the cochlea, and in a few neuronal cell populations. The *POU4F3* transcription factor is required for maturation, survival, and migration of inner and outer hair cells. The mutation results in a nonfunctional protein, which is likely to have a dominant-negative effect in SNHI.
Diagnosis	This syndrome is characterized by late-onset SNHI at the high frequencies, progressing to moderate to severe SNHI at all frequencies. Mutation analysis may confirm the diagnosis.

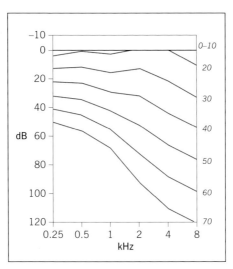

Figure 11. Typical age-related audiogram of the DFNA15 family. Numbers in italics refer to ages (years). Figure courtesy of P Huygen.

Counseling issues Full penetrance has been reported over the age of 40 years. There are no specific recommendations.

DFNA17

MIM	603622
Clinical features	SNHI presents as a mild, high-frequency impairment. Progression results in moderate to severe downsloping SNHI in the second or third decades, with involvement of all frequencies. Severity varies among patients of a similar age. No vestibular symptoms have been reported.
Age of onset	From about 10 years of age onwards.
Epidemiology	One American family with this condition is known.
Inheritance	Autosomal dominant
Chromosomal location	22q11.2
Genes	*MYH9* (nonmuscle, myosin, heavy chain, type A)

Mutational spectrum   A nucleotide substitution causing an R705H amino-acid substitution has been identified.

Effect of mutation   The mutation is expected to affect the ATPase activity of the myosin motor domain (mechanical dysfunction). Dysfunction of the protein leads to cochleosaccular degeneration (Scheibe degeneration) by an unknown mechanism; loss of cochlear and saccular hair cells, and degeneration of supporting cells and stria vascularis is found, in combination with a variable degree of cochlear and vestibular nerve atrophy. Degeneration is usually most severe in the basal turn. Degeneration of the brainstem auditory nuclei may be found in the end stages of the disease.

Diagnosis   Postlingual, progressive SNHI – beginning in the high frequencies – is characteristic. Mutation analysis may confirm the diagnosis.

Counseling issues   Full penetrance has been reported. In one case, cochlear implantation provided little benefit, possibly as a result of peripheral and central degeneration. Other mutations in this gene are responsible for Fechtner syndrome (macrothrombocytopenia, leukocyte inclusions, SNHI, cataracts, and nephritis; MIM 153640); see also dominantly inherited Alport syndrome (p. 59). Affected persons should be screened for the features of Fechtner syndrome. May–Hegglin anomaly (an autosomal dominant macrothrombocytopenia with leukocyte inclusions; MIM 155100) can also be caused by mutations in the *MYH9* gene.

DFNA22

MIM   606346

Clinical features   Postlingual, progressive SNHI. In patients of around 50 years of age, the hearing impairment varies from moderate to profound and involves all frequencies. There is no vestibular involvement.

Age of onset   Childhood

Epidemiology   This condition is known in one family of Italian origin.

Inheritance   Autosomal dominant

Chromosomal location	6q13

Gene	*MYO6* (myosin VI)

Mutational spectrum	A missense mutation, C442Y, has been reported.

Effect of mutation	Myosin VI is a motor molecule (*cf*, MYO7A [Usher syndrome type IB, p. 112 and DFNA11, p. 31] and MYO15 [DFNB3, p. 44]). The mutation affects a conserved amino acid in the motor domain of the molecule and is predicted to destabilize the protein, leading to partial or complete loss of function. SNHI might be caused by haploinsufficiency or by a dominant-negative effect of the mutated protein. In mice, the *myo6* gene is expressed in the inner and outer hair cells of the cochlea and the protein is located at the base of the stereocilia. A mutation in *myo6* leads to fusion of the stereocilia and finally to the death of the hair cells and supporting cells.

Diagnosis	The clinical features together with mutation analysis will provide the diagnosis.

Counseling issues	There are no specific counseling issues.

DFNA36

MIM	606705

Clinical features	Symmetric SNHI with rapid progression occurs within 10–15 years of onset and gives rise to profound SNHI. No vestibular symptoms have been found.

Age of onset	Approximately 5–10 years of age

Epidemiology	One American family with this condition is known.

Inheritance	Autosomal dominant

Chromosomal location	9q13–q21

Genes	*TMC1* (transmembrane cochlear-expressed gene 1)

Mutational spectrum	A nucleotide change resulting in D572N has been described.
Effect of mutation	The mutation probably acts through a dominant-negative mechanism. In mice, the gene is expressed in hair cells of the cochlea and vestibular end organs. The transmembrane protein is required for normal functioning of the cochlear hair cells; however, the exact function remains unknown. In mice heterozygous for a *tmc1* mutation, postnatal cochlear hair cell degeneration is evident.
Diagnosis	This condition is characterized by postlingual, extremely progressive SNHI. Mutation analysis may confirm the diagnosis.
Counseling issues	Full penetrance has been reported. Other mutations in this gene result in recessive SNHI, ie, DFNB7/DFNB11 (see p. 46).

DFNA39

MIM	605594
Clinical features	Mutations in the *DSPP* (dentin sialophosphoprotein) gene give rise to dentinogenesis imperfecta 1, which results in abnormal dentin production and mineralization. Patients with this condition often have opalescent teeth with a blue–gray or amber–brown color. In addition, some mutations in this gene cause SNHI, thus making the disease syndromic. In these cases, DFNA is actually a misnomer, as "DFNA" is defined as being nonsyndromic. All those affected show mild, progressive SNHI starting at the high frequencies. After the age of 50 years, moderate to severe high-frequency SNHI is characteristic. Tinnitus and vestibular symptoms have been reported.
Age of onset	SNHI onset occurs at 20–30 years of age.
Epidemiology	This condition is known in two Chinese families.
Inheritance	Autosomal dominant
Chromosomal location	4q21.3
Genes	*DSPP* (dentin sialophosphoprotein)

Mutational spectrum	Two missense mutations – Pro17Thr and Val18Phe – have been identified.
Effect of mutation	The *DSPP* gene encodes two proteins, dentin sialoprotein (DSP) and dentin phosphoprotein (DPP). The gene is mainly expressed in odontoblasts and transiently in preameloblasts, and also in the inner ear. The mutations probably affect only DSP, which is thought to play an important role in dentinogenesis. The primary effect of the amino-acid substitutions is unknown, but they could interfere with signal peptide cleavage of the DSPP precursor protein.
Diagnosis	This condition is characterized by late-onset SNHI with dentinogenesis imperfecta. The disease should be differentiated from osteogenesis imperfecta types IB and IVB (see p. 90 and p. 95); no bone fractures are found in DFNA39. Mutation analysis can confirm the diagnosis.
Counseling issues	A dentist should evaluate the patient, but there are no further recommendations. Other mutations in the *DSPP* gene result in dentinogenesis imperfecta without SNHI.

Otosclerosis

Otosclerosis is caused by abnormal bone homeostasis of the otic capsule, leading to isolated endochondral bone spongiosis. These "otosclerotic" foci may invade the oval window and stapedial footplate, causing immobility of the latter and thereby leading to conductive or mixed hearing impairment. Pure SNHI rarely occurs.

Otosclerosis has a prevalence of approximately 0.2%–1% in Caucasians and occurs twice as often in women as in men. Onset is usually between the ages of 20 and 40 years, and 90% of patients are less than 50 years old at the time of diagnosis. Both ears are affected in about 85% of cases. The disease is thought to be most often inherited in a dominant fashion, but with low penetrance (25%–50%). SNHI is profound in approximately 10% of cases; this may reflect the presence of cochlear otosclerotic foci. Vestibular dysfunction is found in some patients. Excessive progression has been noted in pregnant women. To date, gene linkage studies have identified three loci: OTSC1, OTSC2, and OTSC3.

OTSC1

MIM	166800
Clinical features	Conductive hearing impairment, ranging from approximately 20 to 45 dB. Age-related progression of impairment may be found as a result of superimposed SNHI.
Epidemiology	One Indian family with 14 affected individuals is known.
Inheritance	Autosomal dominant
Chromosomal location	15q26.1–qter
Genes	Unknown
Mutational spectrum	Unknown
Effect of mutation	Unknown
Diagnosis	Autosomal dominantly inherited otosclerosis is characteristic of this condition. In large families, linkage analysis may confirm the diagnosis.
Counseling issues	Audiograms and stapedial reflexes should be obtained. Stapedotomy may be very successful in restoring hearing. The risk of hearing loss after surgery is generally accepted as less than 1%. In cases with progression of sensorineural thresholds, a CT scan of the cochlea may show cochlear otosclerosis. In these cases, fluoride can be given in an attempt to stop the progression; it seems that a treatment period of about 1–2 years is sufficient. Profound deafness may be caused by severe mixed hearing impairment – in approximately one in three of these patients a stapedotomy may achieve satisfactory air conduction thresholds, which can be rehabilitated with conventional air conduction hearing aids. Patients who are profoundly deaf as a result of otosclerosis, and in whom stapedotomy has failed, may benefit from a cochlear implant. Implantation might be complicated by ossification in the basal turn. Facial nerve stimulation by the implant electrode is frequent in these patients. Penetrance is probably reduced, and thus not all individuals with the mutated allele (chromosome) will develop the disease.

OTSC2

MIM	605727
Clinical features	Conductive hearing impairment.
Epidemiology	One Belgian family with 18 affected members is known.
Inheritance	Autosomal dominant, probably with some reduced penetrance.
Chromosomal location	7q34–q36
Genes	Unknown
Mutational spectrum	Unknown
Effect of mutation	Unknown
Diagnosis	See OTSC1
Counseling issues	See OTSC1

OTSC3

MIM	Not listed
Clinical features	Conductive hearing impairment.
Epidemiology	One Cypriot family with around 10 affected members is known.
Inheritance	Autosomal dominant, probably with some reduced penetrance.
Chromosomal location	6p21.3–p22.3
Genes	Unknown
Mutational spectrum	Unknown
Effect of mutation	Unknown
Diagnosis	See OTSC1
Counseling issues	See OTSC1

Inherited Nonsyndromic Hearing Impairment: Recessive Inheritance

DFNB1

MIM	220290
Clinical features	SNHI with a variable phenotype. The hearing impairment can vary from mild to profound, but is moderate to profound or profound in the majority of cases. Audiograms are either flat or sloping with preferential loss at high frequencies. Progression of the hearing impairment has been reported in a minority of the cases in which this syndrome has been analyzed. Normal vestibular responses have been reported in children.
Age of onset	Prelingual
Epidemiology	In Caucasians who show prelingual SNHI, a mutation in this gene is causative in up to approximately 60% of familial cases and 30% of apparently sporadic cases.
Inheritance	Recessive
Chromosomal location	13q11–q12
Genes	*GJB2* (gap-junction protein β–2; also known as *CX26* [connexin 26])
Mutational spectrum	Amino-acid substitutions, nonsense and splice-site mutations, deletions, and insertions have been found. These lead to premature protein truncation. In-frame deletions have also been reported.
	In patients of Caucasian origin, the 35delG mutation is by far the most frequent. The carrier frequency of the 35delG mutation is highest in southern Europe at 1:35. In northern/middle Europe, North America, and Australia, the frequency is 1:79. The 167delT mutation is most common among the Ashkenazi Jewish population, while the 235delC mutation is most common in Japanese and Korean individuals.
Effect of mutation	Mutations lead to either premature truncation of the protein or amino-acid substitutions. The GJB2 protein forms homomeric or heteromeric gap junctions, which are important for the exchange of ions and small

molecules between cells, and is thought to play a role in the recycling of K^+ from the hair cells to the stria vascularis. Mutations in *GJB2* lead to cochlear dysfunction and cell death.

Diagnosis

The type and onset of hearing impairment may indicate this syndrome, but mutation analysis is the main method of diagnosis. Any SNHI that is recessively inherited or occurs during early childhood should be tested for *GJB2* involvement.

Counseling issues

There is intrafamilial phenotypic variation. Genetic counseling should be offered because of the high carrier frequency of the 35delG mutation in Caucasians and specific mutations in other populations. Provided the hearing impairment is severe to profound, cochlear implantation can be carried out. A normal implantation risk for complications is found. Relatively successful results have been reported for cochlear implantation in DFNB1 patients.

Digenic inheritance has been suggested in some individuals: a combination of a mutation in *GJB2* in one allele and a deletion encompassing part of the *GJB6* gene (also known as *CX30*, flanking *GJB2*) in the second allele. Studies suggest that digenic inheritance might be a relatively common cause of early-onset severe to profound hearing impairment in southern Europe, but not in Caucasians in general. Homozygous deletions in *GJB6* also lead to severe to profound early-onset hearing impairment, demonstrating that at least two genes in the DFNB1 locus can cause SNHI. Erythrokeratoderma variabilis and palmoplantar keratoderma are examples of the involvement of different organs by a single *GJB2* mutation. Some mutations exhibit dominant inheritance (see DFNA3, p. 19).

DFNB2

MIM

600060

Clinical features

Severe to profound SNHI. Some patients have symptoms of vertigo. Vestibular function can be reduced or absent. Carriers (tested in a Chinese Arg244Pro family) may show subclinical decreased vestibular function, and Bekesy audiometry may reveal minor abnormalities (semidominant effect).

Age of onset	Congenital onset was found in two families (both Chinese), and onset between birth and 16 years of age was reported in another family (Tunisian).
Epidemiology	This condition has been reported in one Tunisian family and in two Chinese families out of eight that were evaluated.
Inheritance	Autosomal recessive
Chromosomal location	11q13.5
Genes	*MYO7A* (myosin VIIA)
Mutational spectrum	Two nucleotide substitutions that lead to an amino-acid change or exon skipping have been described, along with a small insertion that leads to protein truncation.
Effect of mutation	Mutations lead to dysfunction of the nonmuscle myosin VIIA. Myosin VIIA is an actin-based motor protein that plays a role in the organization of the inner-ear hair cell stereocilia and in the epithelium of the retinal pigment. The myosin molecule functions in the distribution of melanosomes and in opsin transport in photoreceptor cells.
Diagnosis	Profound SNHI, often with vestibular dysfunction, characterizes this condition. Mutation analysis may confirm the diagnosis.
Counseling issues	Patients should be screened for pigmentary retinopathy. Penetrance might be incomplete and this should be taken into consideration during genetic counseling. Other mutations may cause DFNA11 (see p. 31) or Usher syndrome type IB (see p. 112).

DFNB3

MIM	600316
Clinical features	Profound SNHI affecting all frequencies, usually without vestibular symptoms. Some patients, however, have vestibular signs on the tandem walk and Romberg tests. Heterozygotes in a family from Bali had either normal or borderline-normal hearing.

Age of onset	Congenital
Epidemiology	The SNHI of about 45 inhabitants of the village of Bengkala on Bali and of several unrelated Indian and Pakistani families is caused by mutations at the DFNB3 locus. Among consanguineous families in Pakistan, at least 5% of individuals with congenital profound SNHI have inherited a homozygous DFNB3 mutation.
Inheritance	Autosomal recessive
Chromosomal location	17p11.2
Genes	*MYO15A* (unconventional myosin XVA)
Mutational spectrum	Missense, nonsense, and splice-site mutations have all been described. These lead to protein truncation.
Effect of mutation	Myosins are proteins that bind to actin filaments and function as molecular motors using ATP. The *MYO15A* gene is expressed in the inner and outer hair cells of the cochlea. The protein is mainly present in the stereocilia and the apical cell body. Together with short stereocilia, a changed actin filament organization is found in *myo15*-defective mice, which suggests that the role of myosin XVA in the regulation of actin polymerization is disturbed by the mutations.
Diagnosis	Congenital profound SNHI without obvious vestibular dysfunction suggests the diagnosis. This may be confirmed by mutation analysis.
Counseling issues	There are no specific recommendations. Penetrance is thought to be complete.

DFNB4

(see: EVA syndrome, p. 103)

DFNB7/DFNB11

MIM	600974
Clinical features	Profound SNHI
Age of onset	Congenital
Epidemiology	Out of approximately 230 families with severe to profound SNHI that were screened, this disease was found in 11 Indian and Pakistani families. One additional Indian family and a Bedouin family from Israel have also been identified.
Inheritance	Autosomal recessive
Chromosomal location	9q21.12
Genes	*TMC1* (transmembrane cochlear-expressed gene 1)
Mutational spectrum	Nucleotide substitutions causing missense, nonsense, or splice-site mutations have been found. In addition, a deletion of one nucleotide leading to premature protein truncation and a large deletion encompassing part of the gene have been described.
Effect of mutation	The transmembrane protein is required for the normal function of cochlear hair cells; the mutations disturb this function. In mice, mutations lead to hair cell degeneration. However, the exact function of the protein is unknown.
Diagnosis	Congenital profound SNHI suggests a diagnosis of DFNB7/DFNB11. This can be confirmed by mutation analysis.
Counseling issues	Full penetrance has been reported. Other mutations result in dominant SNHI, eg, DFNA36 (see p. 37).

DFNB8/DFNB10

MIM	601072/605316
Clinical features	In most families, SNHI is congenital and severe to profound – though nonprogressive. One Pakistani family, however, has been described with

extremely progressive SNHI, which, after onset, leads to profound SNHI (up to 105 dB) within 4–5 years.

Age of onset Primarily congenital. Onset has also been described between 0 and 12 years of age (Pakistani family). The age of onset may depend on the specific mutation.

Epidemiology This type of hearing impairment seems to be relatively frequent in populations in which consanguineous marriages are common. Most of the known families are from Pakistan, Tunisia, or Palestine.

Inheritance Autosomal recessive

Chromosomal location 21q22.3

Genes *TMPRSS3* (transmembrane protease, serine 3)

Mutational spectrum Nucleotide substitutions, a small deletion, and a complex rearrangement have been found. These lead to amino-acid substitutions, splicing defects, and frame-shift mutations, causing protein truncation.

Effect of mutation The gene is expressed in several tissues – including the fetal cochlea – and may be involved in development and maintenance of the inner ear. *TMPRSS3* is thought to encode a transmembrane serine protease. Some of the missense mutations are predicted to impair the putative enzymatic activity of the gene. How the defects lead to hearing impairment is unknown. The milder phenotype observed in the Pakistani family may result from a mutation in the splice-acceptor site.

Diagnosis Profound SNHI should raise suspicion for the diagnosis, which may be confirmed by mutation analysis.

Counseling issues None

DFNB9

MIM 601071

Clinical features Severe to profound SNHI affecting all frequencies, without vestibular symptoms.

Age of onset Before the age of 2 years and probably congenital

Epidemiology	This disease has been described in families from Lebanon, Turkey, the United Arab Emirates, and India, and in a Druze family. Importantly, this form of hearing impairment accounts for about 3% of recessively inherited SNHI in Spain.
Inheritance	Autosomal recessive
Chromosomal location	2p22-p23
Genes	*OTOF* (otoferlin)
Mutational spectrum	Nucleotide substitutions causing either missense or nonsense mutations have been found, in addition to a splice-site mutation.
Effect of mutation	It is unknown how the mutations lead to hearing impairment. Otoferlin is predicted to be involved in vesicle membrane fusion. Since the gene is expressed in the inner-ear hair cells, these vesicles may be synaptic vesicles.
Diagnosis	Prelingual severe to profound SNHI suggests the syndrome. Mutation analysis may lead to the diagnosis.
Counseling issues	There are no specific recommendations. Penetrance is complete.

DFNB12

MIM	601386
Clinical features	Moderate to profound SNHI, which can be progressive, without symptoms of vestibular dysfunction. All frequencies are involved.
Age of onset	Congenital
Epidemiology	Families from Europe, Syria, and India, and of African American origin have been described. The relative importance of *CDH23* mutations (see below) as a cause of SNHI in different populations is unknown.
Inheritance	Autosomal recessive
Chromosomal location	10q22.1

Genes	CDH23 (cadherin-related 23)
Mutational spectrum	Only missense mutations have been described for DFNB12. Nonsense, splice-site, and missense mutations in CDH23 have been described to cause Usher syndrome type ID (see p. 112).
Effect of mutation	CDH23 is expressed in both the cochlea and the retina. In the Waltzer mouse, which has mutations in cdh23, disorganization of hair cell stereocilia is reported, resulting in SNHI. It has been hypothesized that CDH23 functions in linking the stereocilia. The majority of the DFNB12 missense mutations are predicted to impair calcium-binding of the protein, thereby impairing proper folding of the extracellular part of the protein.
Diagnosis	The diagnosis is suggested by congenital, moderate to profound SNHI. Mutation analysis may confirm the diagnosis.
Counseling issues	Patients should be monitored for pigmentary retinopathy by an ophthalmologist and for vestibular impairment. Asymptomatic pigmentary retinopathy-like manifestations can be found. Also mutations in CDH23 can lead to a wide range of hearing impairment and pigmentary retinopathy phenotypes, with or without vestibular areflexia.

DFNB21

MIM	603629
Clinical features	Severe to profound SNHI at all frequencies. Heterozygotes seem to have normal hearing.
Age of onset	Prelingual
Epidemiology	This has been found in one Shi'ite Lebanese family.
Inheritance	Autosomal recessive
Chromosomal location	11q22–q24
Genes	TECTA (α-tectorin)
Mutational spectrum	There is a G $\rightarrow$ A transition in the donor splice site of intron 9.

Effect of mutation	The recessively inherited mutation causes truncation of the encoded protein. Alpha-tectorin is one of the major components of the tectorial membrane, which is needed for deflection of the stereocilia of the hair cells. This deflection causes depolarization of the hair cells, which finally results in the perception of sound. Mutations lead to a disturbance in the structure of the tectorial membrane, thereby causing SNHI.
Diagnosis	The diagnosis is suggested by prelingual severe to profound SNHI. Mutation analysis may confirm the diagnosis.
Counseling issues	Other mutations in the *TECTA* gene cause DFNA12 (see p. 25).

DFNB29

MIM	605608
Clinical features	Severe to profound SNHI without symptoms of vestibular dysfunction. This has been confirmed by vestibular tests in some individuals.
Age of onset	Congenital
Epidemiology	Among 100 Pakistani families with recessive nonsyndromic SNHI, two were of the DFNB29 type.
Inheritance	Autosomal recessive
Chromosomal location	21q22.3
Genes	*CLDN14* (claudin 14)
Mutational spectrum	A single nucleotide deletion leads to premature protein truncation, and a nucleotide change causes substitution of a conserved amino acid.
Effect of mutation	The hearing impairment is caused by either a dominant-negative effect of the mutations or haploinsufficiency. In mice, the gene is expressed in the organ of Corti and the sensory epithelia of the vestibular organs. Claudins are components of tight junctions. The transmembrane protein claudin 14 is likely to function in cochlear tight junctions, which modulate paracellular permeability between extracellular compartments (separation of scala media from other compartments).

Diagnosis	Congenital severe to profound SNHI suggests the diagnosis, which may be confirmed by mutation analysis.
Counseling issues	None

Inherited Nonsyndromic Hearing Impairment: X-linked Inheritance

DFN1

(see: Mohr Tranebjaerg syndrome, p. 78)

DFN3

(also known as: X-linked mixed deafness with stapes fixation and perilymphatic gusher)

MIM	304400
Clinical features	Hearing impairment is of a mixed type, consisting of conductive and sensorineural components. In males, hearing impairment starts with the conductive component in the low frequencies. There is almost no conductive hearing impairment in the high frequencies. SNHI progresses over time to severe or profound. Vestibular anomalies can be present. In cases with a 30–40 dB conductive component, this could be due to stapes fixation. If the conductive hearing impairment is milder, this is considered to be due to a widened vestibule. High-resolution CT shows the temporal bone defects: dilatation of the vestibule and a dilated internal auditory canal (see **Figure 12**). Perilymphatic gusher refers to the heavy flow of cerebrospinal fluid after the stapedial footplate has been surgically opened to replace the stapes. Females (with a heterozygous mutation) can suffer from mixed hearing impairment. So far, no CT-detectable anomalies have been reported for female carriers.
Age of onset	Early childhood
Epidemiology	DFN3 is the most frequent X-linked form of hereditary hearing impairment – it appears to account for more than 50% of cases of X-linked hearing impairment.

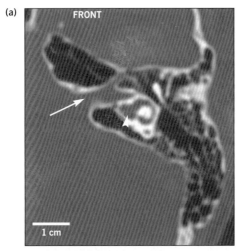

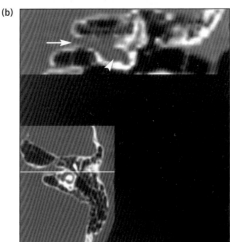

Figure 12. (a) Axial CT scan of the left ear of a patient with DFN3. The arrow indicates the internal auditory canal, which shows lateral dilatation. The vestibule is also enlarged (arrowhead). **(b)** A reformatted coronal CT scan of the same ear. The internal auditory canal is again indicated with an arrow. The arrowhead indicates the position of the inner ear, which seems to communicate with the internal auditory canal. This probably results in conduction of cerebrospinal fluid pressure to the cochlea, thereby causing hearing impairment and perilymphatic gusher during surgery.

Inheritance	X-linked
Chromosomal location	Xq21.1
Genes	*POU3F4* (POU domain, class 3, transcription factor 4)
Mutational spectrum	Missense and nonsense mutations, a small in-frame deletion (two amino acids), and deletions encompassing the entire gene have been described. In addition, deletions 900-kb upstream of the gene have been described, encompassing an 8-kb critical segment and a deletion/inversion and duplication/inversion involving these sequences. One deletion involving sequences between the gene and the 8-kb segment has been found.
Effect of mutation	The mutations lead to a nonfunctional protein or cause its dysfunction. The *POU3F4* gene encodes a transcription factor and thus regulates the activity of a number of other genes. This regulation is thought to be disturbed by the mutations. Most missense mutations probably impair the binding of *POU3F4* to its target DNA sequences in genes regulated by *POU3F4*. The deletions upstream of the gene appear to affect its expression.
	In mice, the *pou3f4* gene is active very early in the development of the otic capsule. From studies in mouse models with a mutated *pou3f4* gene, it can be concluded that POU3F4 enhances the survival of mesodermal cells in mesenchyme remodeling. Mesenchyme remodeling forms the bony labyrinth and regulates inductive signaling in the otic mesenchyme. Mutant mice show a variable phenotype, with hearing impairment and vestibular dysfunction. Several anatomic and histologic abnormalities are described in these mice, including abnormalities in the middle-ear ossicles, a reduction in the coiling of the cochlea, and hypoplasia of otic mesenchyme-derived cochlear structures such as the spiral limbus, scala tympani, and strial fibrocytes. In addition, there is a bulbous dilatation of the lateral portion of the internal auditory canal with incomplete separation from the cochlea and widening of the first part of the facial nerve.
Diagnosis	A mixed type of hearing impairment on audiologic testing combined with detailed CT scanning (males) leads to the diagnosis. Family data showing X-linked inheritance and mutation analysis can confirm this.

A mutation is found in more than 80% of the radiologically diagnosed cases by analysis of the coding region and the critical 8-kb upstream region, 900-kb proximal to *POU3F4*.

Counseling issues   Perilymphatic gusher may complicate stapes surgery in these patients. This can cause dizziness and lead to increased hearing impairment. Therefore, the diagnosis is of crucial importance before performing stapes surgery. In one case, a mosaic *de novo* mutation was found. The mutation was shown to be present in only about 50% of peripheral blood cell DNA. Missense mutations are clustered in the POU homeodomain. Two females have been described with clinical features as severe as those in males. This is probably due to preferential inactivation of the X chromosome with the normal *POU3F4* allele. Missense mutations are clustered in the POU homeodomain.

2. Inherited Syndromic Hearing Impairment

Alport Syndrome

This syndrome may be caused by mutations in at least four genes. Inheritance is predominantly X-linked dominant (85% of cases), with autosomal recessive inheritance occurring in about 10% of cases. Autosomal dominant inheritance is rare. Hearing impairment is found in at least 55% of males and 45% of females in all Alport types.

(a) Alport Syndrome, X-linked

MIM   301050

Clinical features   Progressive nephritis, often in combination with SNHI. Eye abnormalities might be present. SNHI (mostly high frequency) usually appears when renal function deteriorates; this varies from about 10 years of age to late adulthood. In cases with an early onset of hearing impairment, the low frequencies may also be affected. These impairments may necessitate the use of a hearing aid. Some patients never develop SNHI.

The onset and severity of renal disease and SNHI depend upon the effect of the mutation on the normal function of the *COL4A5* gene (see below). A microscopic hematuria may be found from early childhood onwards; this can become macroscopic in adolescence or adulthood due to deteriorating renal function. In families with juvenile Alport syndrome, the mean age at development of end-stage renal disease (ESRD) is <31 years; in the adult type, the age is >31 years. The standard deviation around the mean age (in a specific family) at ESRD is 5–7 years; this knowledge can be used for counseling related affected males.

Ocular symptoms (which occur in about 10% of all Alport patients) not specifically related to a particular type of inheritance consist of anterior lenticonus (in which the lens protrudes into the anterior chamber), perimacular pigmentary changes, and flecks around the foveal area.

In combination with a deletion in the *COL4A6* gene, the syndrome predisposes to the development of multiple leiomyomas involving the esophagus, tracheobronchial tree, and/or genitalia (in women). Males with this combination develop severe early-onset Alport syndrome; females exhibit milder renal and auditory symptoms. In either sex, cataracts may appear with this combined gene defect.

Age of onset	Fully dependent on the type of mutation
Epidemiology	All Alport types together: about 1:200,000
Inheritance	X-linked
Chromosomal location	Xq22.3. There is probably a second locus on the X chromosome.
Genes	COL4A5 (type IV collagen α5 chain), possibly COL4A6
Mutational spectrum	Several types of mutations have been described, including large- and medium-size deletions, in-frame deletions, an in-frame duplication, amino-acid substitutions, nonsense mutations, and splice-site mutations.
Effect of mutation	An abnormal collagen IV α-chain results in defective folding of collagen heterotrimers and degeneration of α-chains. This causes focal thinning and thickening, and eventually leads to basement membrane splitting in the glomerulus. SNHI is probably due to a disturbance in the function of the basilar membrane, spiral ligament, and stria vascularis, in which collagen is expressed in the cochlea.
Diagnosis	Hematuria combined with loss of renal function and high-frequency SNHI or a positive family history for these symptoms suggests this diagnosis. Carriers (heterozygous females) often show high-frequency SNHI. Electron microscopy of a renal biopsy may confirm the disease. Mutation analysis provides a definitive diagnosis.
Counseling issues	Estimated penetrance is about 1.00 in males and 0.85 in females. Males are affected earlier and more severely. Females often only suffer from microhematuria (90% of patients) or are even asymptomatic; however, sensitive audiometric tests of these asymptomatic carriers can show either slight high-frequency or mid-frequency SNHI. Sometimes, women are similarly affected to men, although less than 10%–15% of women develop chronic renal failure before the age of 40–50 years. No treatment is available to prevent the progression of renal disease for any type of Alport syndrome. ESRD can only be treated with dialysis or renal transplantation.

(b) Alport Syndrome, Autosomal Recessive

MIM	203780
Clinical features	Progressive nephritis in combination with SNHI (see X-linked type). Eye abnormalities may occur.
Age of onset	See X-linked type
Epidemiology	See X-linked type
Inheritance	Autosomal recessive
Chromosomal location	2q36–q37
Genes	*COL4A3* (type IV collagen α3 chain), *COL4A4*
Mutational spectrum	Nonsense mutations or a small deletion in *COL4A3* create a premature stop codon. Nonsense, missense, and splice-site mutations are found in *COL4A4*, in addition to small deletions or insertions leading a frame shift.
Effect of mutation	See X-linked type
Diagnosis	Hematuria combined with loss of renal function, high-frequency SNHI, or a positive family history of these symptoms indicates the diagnosis. Renal biopsy for electron microscopy or mutation analysis confirms this disease.
Counseling issues	Males can transmit the mutation to both sons and daughters. Males and females are equally affected. Patients with *COL4A4* defects can have normal hearing thresholds. Usually, a juvenile Alport syndrome is found. The other findings are more or less the same as for the X-linked type.

(c) Alport Syndrome, Autosomal Dominant

MIM	104200
Clinical features	Progressive nephritis, often in combination with SNHI. Mutations in the *MYH9* gene have been found to be responsible for Fechtner/Epstein syndrome (hereditary nephropathy and deafness associated with hematologic abnormalities), causing autosomal dominantly inherited

Alport syndrome with macrothrombocytopenia. In Fechtner syndrome, an additional congenital cataract is found (see also DFNA17, p. 35). It has even been suggested that autosomal dominant Alport syndrome is in fact Fechtner/Epstein syndrome.

Age of onset	Variable, age range unknown
Epidemiology	See X-linked type
Inheritance	Autosomal dominant
Chromosomal location	2q
Genes	*COL4A3* (type IV collagen α3 chain) and/or *COL4A4*
Mutational spectrum	A missense mutation in the *COL4A4* gene, and splice-site mutations and short in-frame deletions in *COL4A3*.
Effect of mutation	See X-linked type
Diagnosis	See recessive type
Counseling issues	Offspring of an affected parent will have a 50% risk of inheriting the disease. Dominantly inherited Alport syndrome can have a relatively mild phenotype, with late-onset SNHI and slow progression of renal disease.

Branchio-oto-renal Syndrome

(also known as: BOR syndrome)

MIM	113650, 601653
Clinical features	Branchial arch anomalies (cup-shaped pinnae, preauricular pits, tags, and second branchial arch fistulae; see **Figures 1** and **2**) with conductive, mixed – though sometimes pure – SNHI, and renal malformations. The middle ear shows underdevelopment or absence of the ossicles, and often stapes fixation. The cochlea and semicircular canals may be absent or underdeveloped, resulting in a Mondini-type malformation. Enlargement of the vestibular aqueduct can be present – fluctuating, progressive SNHI is more frequent in these patients. Renal anomalies (varying from undetected to bilateral aplasia) and other

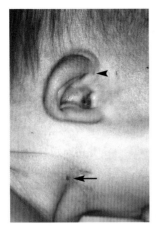

Figure 1. Patient showing a cup ear, which is a typical feature of branchio-oto-renal syndrome. A preauricular pit (arrowhead) and cervical fistula (arrow) are visible. Figure courtesy of H Marres.

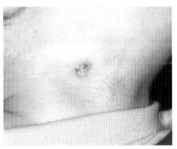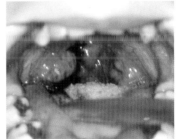

Figure 2. (**a**) A close-up of the cervical fistula. Repetitive infection of a fistula may occur. (**b**) Injection of methylene blue into the fistula reveals its relation with the tonsillar fossa, where the other end of the fistula is situated. Figure courtesy of H Marres.

anomalies of the collecting system can be seen (see **Figure 3**). In addition, cataracts may be found. When renal abnormalities and second branchial arch fistulae are absent, this syndrome is called "branchio-oto (BO) syndrome". If hearing impairment does not occur, these anomalies are referred to as "branchio-renal (BR) anomalies". Gustatory lacrimation may occur subsequent to lacrimal duct aplasia or stenosis.

Age of onset   Congenital

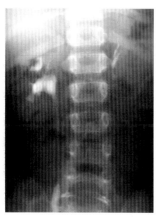

Figure 3. Severe hypoplasia of the left kidney in a patient with branchio-oto-renal syndrome is visualized by an intravenous pyelogram. Figure courtesy of H Marres.

Epidemiology	This syndrome has a prevalence of about 2.5:100,000 persons and accounts for about 2% of profoundly deaf children.
Inheritance	Autosomal dominant
Chromosomal location	8q13.3. A second locus at 1q31 has been identified in two families with BO syndrome.
Genes	*EYA1* (eyes absent 1); the causative gene on 1q31 has not yet been determined.
Mutational spectrum	Nucleotide substitutions leading to amino-acid substitutions or protein truncation, splice-site mutations, small deletions and insertions, gross deletions, larger insertions, and rearrangements.
Effect of mutation	Disruption of inner-ear development (specification and differentiation), the organ of Corti, and sensory epithelia of the vestibular system. *EYA1* is also expressed in metanephric cells. The *EYA* gene is also involved in eye development, hence the formation of cataracts in some individuals.
Diagnosis	Autosomal dominantly inherited branchial arch anomalies and hearing impairment, often with renal malformations, indicate the diagnosis. Mutation analysis of the *EYA1* gene can be performed, although this is not necessarily confirmatory as no mutations in *EYA1* are detected in about 70% of families with the BOR phenotype.

Counseling issues	Variable expression (intra/interfamilial) and almost 100% penetrance have been reported. Branchial manifestations are usually without consequence; however, abscess formation can occur. Hearing impairment is found in about 93% of patients. Hearing impairment is usually of mixed character (45%), although it can also be solely conductive (29%) or solely sensorineural (26%). SNHI may be fluctuating and progressive over time. A mixed or conductive hearing impairment can be partially corrected by ear surgery. The use of a hearing aid is often necessary. Treatment of renal anomalies depends strongly on the type of malformation. Because of considerable variability of expression of the various features, it is difficult to give adequate counseling. It is not known what percentage of families is affected by defects in either or both loci.

CHARGE Association

(also known as: coloboma (C), heart anomaly (H), choanal atresia (A), retardation (R), genitourinary (G) and ear (E) anomalies, included)

MIM	214000
Clinical features	Ocular colobomas (80%–90% of patients) range from iriscolobomas to retinal colobomas, optic nerve colobomas, or clinical anophthalmia. Congenital heart defects (60%–90%) consist of atrial septal defects, ventricular septal defects, patent ductus arteriosus, mitral valve prolapse, coarctation of the aorta, double-outlet right ventricle, pulmonary stenosis, hypoplastic left heart, and tetralogy of Fallot. Choanal atresia, which is found in 55%–65% of patients, is bilateral in 65% and occurs more often on the left side in patients with unilateral atresia. Retardation of growth and/or development occurs in 80%–100% of patients, while 35%–90% have genitourinary abnormalities. Ear anomalies and/or hearing impairment or deafness occur in 80%–95% of patients. External ear anomalies are quite typical and the auricles are described as low-set, short, simple, lop- or cup-shaped, and posteriorly rotated (see **Figure 4**). Patients also show absence or hypoplasia of the semicircular canals (100%), obliteration of the oval window (100%), and variable ossicular chain malformation (66%) (see **Figures 5–7**). Additional features include cleft lip/palate (15%–20%), facial palsy, swallowing difficulties, tracheo-esophageal fistula (15%–20%), central nervous system anomalies

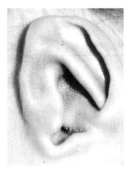

Figure 4. Abnormal appearance of the external ear in CHARGE association. Also note the triangular-shaped concha auriculae.

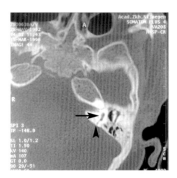

Figure 5. Axial CT scan of the left ear at the level of the internal auditory canal and facial nerve. Both have a normal appearance. The vestibule is small (arrow) and semicircular canals are absent. The vestibular aqueduct is short, but not enlarged (arrowhead). Pneumatization of the mastoid is diminished.

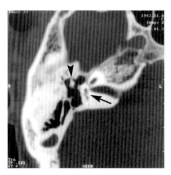

Figure 6. Axial CT scan of the right ear of another patient showing an almost absent vestibule (arrow) and absent semicircular canals. The body of the incus is small (arrowhead).

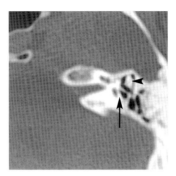

Figure 7. Axial CT scan of the left ear of a third patient showing bony obliteration of the oval window (arrow) and a deformed body of the incus (arrowhead). Pneumatization of the mastoid is diminished.

Figure 8. A 4-year-old girl with CHARGE association. Note the square face, the simple ear shape on the right side, and the low-set left ear with a triangular-shaped concha auriculae. The patient's vision is impaired as a result of a bilateral coloboma of the eye. A CT scan of the temporal bone revealed a bilateral dysplastic cochlea and absent semicircular canals resulting in vestibular areflexia; however, her hearing is normal.

(including cranial nerve dysfunction [55%–90%]), nystagmus (25%), and skeletal abnormalities (10%–15%). A 4-year-old girl with CHARGE association is shown in **Figure 8**.

Age of onset	Congenital
Epidemiology	Prevalence has been estimated at 1:10,000.
Inheritance	Most cases are sporadic, although a few reports mention autosomal dominant and autosomal recessive inheritance.

Chromosomal location	Although the chromosomal location of CHARGE association is not known, numerous chromosomal abnormalities have been found in association with CHARGE-like features in patients with cat-eye syndrome, trisomy 22, deletions in 22q11, triploidy, long-arm deletions of chromosomes 9, 11, and 13, partial duplications of chromosome 4, duplication (14)(q22–q24.3), balanced whole-arm translocation 6;8, translocation 2;18, and translocation 3;22, t(2;7)(p14;q21.11).
Genes	Unknown
Diagnosis	Classical diagnostic criteria include: (a) one or both of the major features (coloboma and choanal atresia); and (b) at least four of the other anomalies designated by the acronym. Vestibular dysfunction and anomalies of the semicircular canals contribute highly to the diagnosis. The latter is a very sensitive and specific finding of CHARGE association.
Counseling issues	Ophthalmologic examination can contribute to the diagnosis, and the assessment of visual function is important to optimize the patient's potential for joining the education system. Congenital heart defects can be life-threatening. A functional airway should be established, sometimes requiring tracheostomy. There are different opinions about the most appropriate time for surgery on the choanal atresia. The diagnosis of mental retardation should be made with caution in patients with combined visual and auditory defects. Delays in motor development are due to vestibular dysfunctions and physical therapy is often required. Hypogonadotropic hypogonadism can be treated by hormone therapy. A hearing aid can alleviate the consequences of mixed conductive and sensorineural deafness. Insertion of grommets for chronic serous otitis media is often necessary. Formation of a feeding gastrostomy is often required due to serious feeding problems.

Craniosynostoses

(a) Apert Syndrome

(also known as: acrocephalosyndactyly type I; Apert Crouzon disease)

MIM	101200

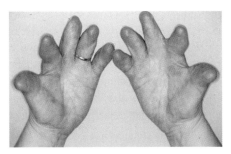

Figure 9. Symmetrical syndactyly of the hands, which may be osseous and/or cutaneous.

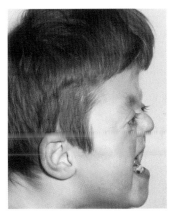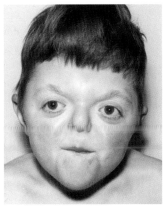

Figure 10. A patient with the typical appearance of Apert syndrome. Note the brachycephaly, flat facies, relative mandibular prognathism, and beaked nose. Hypertelorism and downslanting palpebral fissures are ocular findings in this patient.

Clinical features

Craniosynostosis, midfacial malformations, and syndactyly of the hands and feet (see **Figure 9**). The skull shows brachycephaly, a steep forehead, and flat occiput. Irregular synostosis is observed with late closure of the fontanels. Flat facies, shallow orbits (often with proptosis), maxillary hypoplasia (causing a relative mandibular prognathism), and a small, often beaked nose with septum deviation are observed (see **Figure 10**). An expanded ethmoidal labyrinth, enlarged sella, choanal stenosis or atresia, and short clivus may be found. The oropharynx may show a high-arched palate, sometimes accompanied by a cleft soft palate or bifid uvula (30% of patients). Besides these, an abnormal eruption of the teeth with malocclusion may be seen.

Ocular findings include hypertelorism, strabismus, and downslanting palpebral fissures. Various abnormalities of the external ocular muscles – causing motility disturbances – have been reported. Hearing impairment of a conductive origin caused by otitis media with effusion (common) and/or fixation of the stapedial footplate (rare) may be found. A predisposition for chronic otitis media has been noted. A dehiscent jugular bulb has been sporadically reported. The ears may be low set.

Osseous and/or cutaneous total or partial syndactyly, most commonly of the second, third, and fourth digits, are observed. The feet show cutaneous syndactyly of all toes, with or without osseous syndactyly. Distal phalanges of the thumb and distal hallux may be broad.

Mean birth weight, length, and head circumference may be above the 50th percentile. A deceleration is observed during childhood, leading to values between the 5th and 50th percentile. Deceleration becomes even more pronounced after adolescence, and the mean adult height is also between the 5th and 50th percentile. Besides progressive hydrocephalus, which is an uncommon finding, several other anatomical malformations can occur in the brain. These result in a mean IQ of about 60–75; however, normal intelligence has been reported. Other features of the syndrome are moderate to severe acne vulgaris (70% of patients) and fusion of cervical vertebrae (68%), usually C5–C6. Other limb abnormalities and gastrointestinal, respiratory, genitourinary, and cardiac defects are occasionally found.

Age of onset	Congenital
Epidemiology	The incidence is thought to be 1.5:100,000. This syndrome accounts for about 4.5% of all craniosynostoses.
Inheritance	Autosomal dominant
Chromosomal location	10q26
Genes	*FGFR2* (fibroblast growth factor receptor 2)
Mutational spectrum	Two mutations in exon 7 (IIIa), S252W (934C–G) and P253R (937C–G), are responsible for nearly all cases. In some cases, substitution of a different amino acid occurs at these loci. In two cases, insertion of an Alu element was found in, or close to, exon 9.

Effect of mutation   The mutated amino acids are located in the linker region between immunoglobulin-like domains II and III of the protein. The mutations affect the binding of fibroblast growth factors to the receptor and an activating effect has been indicated. Also, ligand specificity may be decreased. The mutations result in an increased subperiosteal bone-matrix formation. Premature calvaria ossification during fetal development is caused by increased maturation of preosteoblast cells.

Diagnosis   The clinical features may lead to the diagnosis. Mutation analysis can give a definitive diagnosis. Differential diagnoses include nonsyndromic craniosynostosis, Crouzon syndrome, Pfeiffer syndrome, Saethre–Chotzen syndrome, and Jackson–Weiss syndrome (craniosynostosis and broad big toes with medial deviation).

Counseling issues   There may be marked phenotypic variability. Nearly all cases are sporadic. For these cases, the parents of the affected child have a negligible risk of producing another affected child. However, if a parent is affected then the recurrence risk (ie, the risk to a couple with an affected child of having another child with the disease) is about 50%. Most new mutations are of paternal origin and a paternal age effect might be present. The S252W mutation results significantly more often in a cleft palate and the P253R mutation results significantly more often in severe syndactyly of the hands and feet. Early surgery of hand anomalies may be indicated. All patients should have a baseline CT and MRI to trace central nervous system abnormalities. The diagnostic evaluation should also involve a search for cardiovascular (10%) and genitourinary (9.6%) anomalies. Surgery for synostosis is indicated when increased intracranial pressure is found. Mental retardation is usually the result of anatomical abnormalities in the brain; however, skull decompression before the age of 1 year results in a significantly higher IQ in many patients. At birth, abnormalities of the airway can lead to early death. Patients may be operated upon for a conductive hearing impairment. Extensive facial plastic surgery may be performed. The outcome of craniofacial surgery is better in patients with the P253R mutation.

(b) Crouzon Syndrome

(also known as: craniofacial dysostosis; Crouzon craniofacial dysostosis)

MIM	123500
Clinical features	Craniosynostosis and midfacial malformations. Other features that are sometimes observed include brachycephaly due to synostosis of cranial sutures with palpable ridges (see **Figure 11**), hypertelorism, maxillary hypoplasia with relative mandibular prognathism, frontal bossing, parrot-like nose, dental crowding and malocclusion, and calcification of the stylohyoid ligament (88% of patients). About 50% of patients have lateral palatal swellings. Deviation of the nasal septum is found in about 33% of patients and cervical fusion (often C2–C3) in 25%. In 75% of cases, exophthalmus may occur due to shallow orbits (sometimes causing exposure conjunctivitis or keratitis). Exotropia and divergent strabismus are common, as is poor vision (46%); optic atrophy and blindness may be found in these cases.
	The ears may show atresia of the external ear canal (13%), absence of the drum, and abnormalities of the ossicular chain. These features can lead to a conductive hearing impairment (55%). Congenital ossicular chain abnormalities include deformity of the stapes, bony fusion to the promontory, and ankylosis of the malleus/incus to the outer wall of the epitympanum (see **Figure 12**). The round and/or oval window may be narrowed. CT scanning of the temporal bone may show deformity of the internal acoustic canal, outward rotation of the petrous pyramids, an atypical course of the facial nerve, and hyperostosis.
	Progressive hydrocephalus, chronic tonsillar herniation, syringomyelia, torticollis, seizures (12%), headaches (29%), and mental retardation (3%) have also been reported. Sleep apnea occurs in some patients.
Age of onset	The diagnosis can usually be made at birth or during the first year of life. Craniosynostosis begins very early and is often completed by the age of 2–3 years. Sporadically, characteristics are absent at birth and only develop during the first few years of life.
Epidemiology	The incidence is around 1.6:100,000 births. This syndrome accounts for approximately 4.5% of all craniosynostoses.

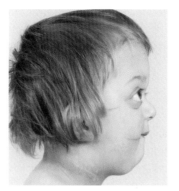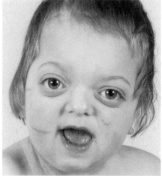

Figure 11. A young girl (about 5 years) with Crouzon syndrome. Hypertelorism, exophthalmus, and brachycephaly are present

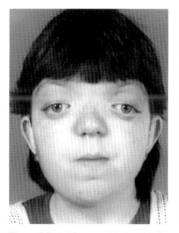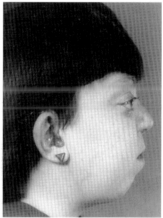

Figure 12. A 12-year-old girl with Crouzon syndrome. Note hypertelorism, exophthalmus, and hypoplasia of the maxilla. Bilateral partial atresia of the external ear canal and fixation of the malleus, incus, and stapes were found. The facial nerve on the left side almost completely covered the oval window.

Inheritance	Autosomal dominant
Chromosomal location	10q26
Genes	FGFR2 (fibroblast growth factor receptor 2)

Mutational spectrum	Missense mutations are mainly found; occasionally there are small in-frame deletions or insertions. The mutations are located in the immunoglobulin III domain and adjacent linker regions of the protein (exons IIIa and IIIc).
Effect of mutation	Many of the mutations lead to the replacement of a cysteine or to the substitution of a cysteine for other amino acids in the protein, which has been predicted to lead to a constitutively activated receptor. The mutations lead to delayed regional skeletal maturation.
Diagnosis	The clinical features point to the diagnosis, which can be confirmed by mutation analysis. The disease-causing mutation is found in about 50% of patients. Differential diagnoses include nonsyndromic craniosynostosis, Apert syndrome, Pfeiffer syndrome, Saethre–Chotzen syndrome, and Jackson–Weiss syndrome (craniosynostosis and broad big toes with medial deviation).
Counseling issues	Incomplete penetrance has been reported (44%–67%). The disease is often sporadic (33%–56%). In sporadic cases, a paternal age effect may be present. Intra- and interfamilial variability are seen. Progressive hydrocephalus with increased intracranial pressure needing surgical treatment might be present. Deviation of the nasal septum can be corrected surgically. Malocclusion can be corrected by orthodontics and, at a later stage, by maxillary advancement. Other operation indications are usually for cosmetic reasons. Surgery for fronto-orbital midfacial advancement, presently by means of gradual distraction, can be performed. Crouzon syndrome in combination with acanthosis nigricans results from a specific mutation: Ala391Glu in the *FGFR3* gene. Other mutations in *FGFR3* lead to achondroplasia, hypochondroplasia, and thanatophoric dwarfism.

(c) Pfeiffer Syndrome

(also known as: acrocephalosyndactyly type V)

MIM	101600
Clinical features	Craniosynostosis, midfacial malformations, mild soft-tissue syndactyly, and broad thumbs and broad big toes. A turribrachycephaly with craniosynostosis of at least the coronal suture, and a steep forehead are

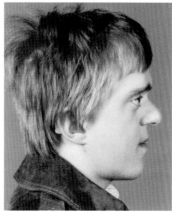

Figure 13. A 14-year-old patient suffering from Pfeiffer syndrome. Note the brachycephaly, steep forehead, proptosis, hypertelorism, midfacial hypoplasia, and beaked nose. This patient also suffered from bilateral conductive hearing impairment.

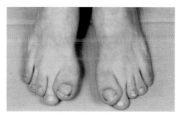
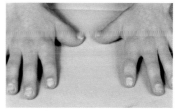

Figure 14. The hands and feet of the same patient. The main abnormalities are the appearance and position of the thumbs and big toes.

found, and a Kleblattschädel (cloverleaf skull) has been described in some cases (see **Figure 13**). A low nasal bridge, long philtrum, small nose (often beaked and narrow), and hypoplastic maxilla with relative mandibular prognathism are usually found. Choanal atresia has occasionally been reported. The oropharynx shows a high-arched palate and crowded teeth.

Ocular findings comprise hypertelorism, proptosis, downslanting palpebral fissures, and strabismus. Conductive hearing impairment has been noted in some patients. Ossicular chain fixation consists of fusion of the incus to the epitympanum and ankylosis of the stapes. CT scanning may reveal stenosis/atresia of the external ear canals, hypoplasia or enlargement of the middle-ear cavity, and hypoplastic ossicles.

The hands and feet show a broad distal phalanx with a varus deformity (see **Figure 14**). A partial soft-tissue syndactyly of the second and third fingers, and of the second, third, and fourth toes, is found. Sporadically, a cartilaginous trachea or laryngo-, tracheo-, or bronchomalacia is found – this can result in severe respiratory distress. Other sporadic findings include fused vertebrae, Arnold–Chiari malformation, mental retardation, hydrocephalus, and seizures.

Age of onset	Congenital
Epidemiology	Unknown
Inheritance	Autosomal dominant
Chromosomal location	8p12, 10q26
Genes	*FGFR1* (fibroblast growth factor receptor 1), *FGFR2*
Mutational spectrum	The *FGFR1* gene contains missense mutations. The *FGFR2* gene contains mainly missense mutations, but also some intron mutations affecting the exons encoding the immunoglobulin III domain and the flanking linker regions.
Effect of mutation	See Crouzon syndrome
Diagnosis	The clinical features indicate the diagnosis, which can be confirmed by mutation analysis. Differential diagnoses include nonsyndromic craniosynostosis, Apert syndrome, Crouzon syndrome, Saethre–Chotzen syndrome, and Jackson–Weiss syndrome (craniosynostosis and broad big toes with medial deviation).
Counseling issues	Three subtypes are distinguished according to prognostic significance. Type 1, showing the classic phenotype, is compatible with life. Types 2 and 3 (with severe neurologic pathology) show a more severe phenotype, and both types lead to death within the first years of life. The disorder is sporadic in many cases – all type 2 and type 3 patients seem to be sporadic. A paternal age effect may be present in these cases. Penetrance is complete and intra- and interfamilial expression is variable.

Figure 15. A young male exhibiting Saethre–Chotzen syndrome. A craniotomy was performed at the age of 8 months because of premature closure of cranial sutures. In addition, abnormalities of the teeth and a bilateral conductive hearing impairment were found. Note hypertelorism, mild ptosis, and a low frontal hairline. Figure courtesy of R Ensink.

(d) Saethre–Chotzen Syndrome

(also known as: Chotzen syndrome; acrocephalosyndactyly type III)

MIM	101400
Clinical features	Brachycephaly, maxillary hypoplasia, prominent ear crus, and mild syndactyly. Brachycephaly with craniosynostosis, late-closing fontanels, a steep forehead, and low-set frontal hairline have been reported (see **Figure 15**). Facial asymmetry with maxillary hypoplasia resulting in relative mandibular prognathism, straight nasofrontal angle, beaked nose, and nasal septum deviation may be found. The oropharynx shows a highly arched palate, malocclusion, supernumerary teeth, and (sporadically) a cleft palate.

Ocular findings include shallow orbits, hypertelorism, ptosis of the eyelid, strabismus, and lacrimal duct anomalies. The ears show a prominent crus, are small, and might be posteriorly angulated. A mild to moderate, usually conductive, hearing impairment is found in about 15% of patients.

Soft-tissue syndactyly of the second and third fingers (see **Figure 16**) and the third and fourth toes may be found, along with various skeletal anomalies. Occasionally, increased intracranial pressure, mental

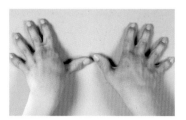

Figure 16. Cutaneous syndactyly of the second and third fingers. Figure courtesy of R Ensink.

deficiency, profound SNHI, short stature, and renal and cardiovascular defects have been found.

Age of onset	Congenital
Epidemiology	This is one of the most common craniosynostoses. Incidence is around 2–4:100,000 births.
Inheritance	Autosomal dominant
Chromosomal location	7p21–p22 (*TWIST*), 10q26 (*FGFR2*), 4p16.3 (*FGFR3*)
Genes	*TWIST*, *FGFR2* (fibroblast growth factor receptor 2), *FGFR3*
Mutational spectrum	The TWIST protein is an upstream regulator of the FGFRs. Nonsense and frame-shift mutations are often found in the *TWIST* gene, leading to premature protein termination. Missense mutations are found in the DNA-binding domain and the helix–loop–helix domain. In the latter, seven amino-acid duplications can be present. In addition to this, cytogenetically visible deletions that encompass the gene are seen, as are translocations with breakpoints 3′ of the gene. The latter are likely to affect regulatory sequences. For the *FGFR2* gene, an amino-acid substitution and a deletion of two amino acids in the linker region between the IgII and IgIII immunoglobulin-like domains have been found. An amino-acid substitution is seen in *FGFR3*.
Effect of mutation	The large deletions suggest that loss of TWIST function causes Saethre–Chotzen syndrome; *TWIST* mutations have been described to promote osteoblast apoptosis and increase the osteogenic capabilities of osteoblastic cells.

Diagnosis	The clinical features may lead to the diagnosis, which can be confirmed by mutation analysis. Differential diagnoses include nonsyndromic craniosynostosis, Apert syndrome, Crouzon syndrome, and Jackson–Weiss syndrome (craniosynostosis and broad big toes with medial deviation).
Counseling issues	Penetrance is high. There is intra- and interfamilial variability in expression of the disease. The majority of patients have a mutation in the *TWIST* gene. Patients with a large deletion encompassing *TWIST* are more likely to have mild to moderate mental retardation.

Jervell and Lange-Nielsen Syndrome

MIM	220400
Clinical features	Prolongation of the QT interval, ventricular arrhythmias, syncopal episodes, and severe to profound perceptive hearing impairment. SNHI usually involves the high frequencies. Arrhythmias not only cause syncopes, but can also be responsible for seizures and sudden death, which can be stress induced. One patient was even reported to die after hearing the sound of an alarm clock.
Age of onset	Cardiac arrhythmias occur earlier in males than in females, generally at puberty. SNHI is usually congenital.
Epidemiology	About 1% of children with prelingual SNHI suffer from Jervell and Lange-Nielsen syndrome.
Inheritance	Autosomal recessive
Chromosomal location	11p15.5, 21q22.1–q22.2
Genes	*KCNQ1* (K$^+$ channel, voltage-gated, KQT-like subfamily, member 1; also known as *KVLQT1*), *KCNE1* (K$^+$ channel, voltage-gated, ISK-related subfamily, member 1)
Mutational spectrum	Most mutations in the *KCNQ1* gene are nonsense or frame-shift mutations leading to premature truncation of the protein. Some amino-acid substitutions have been reported. The mutations in *KCNE1* are missense mutations.

Effect of mutation   The proteins derived from the mutated *KCNQ1* gene are predicted to be unable to assemble into channels with either the ISK protein (*KCNE1* gene) or themselves. Alternatively, the truncated proteins might be unstable and degrade immediately after synthesis. The effect of the missense mutations in the *KCNE1* gene is unknown. The heteromeric K^+ channel controls the ventricular repolarization process. In the inner ear (marginal cells of the stria vascularis and dark cells), the K^+ channel controls endolymph homeostasis.

Diagnosis   The QT interval cannot be used as a basis for diagnosis. Mutation analysis of *KCNQ1* and/or *KCNE1* can be performed.

Counseling issues   The *KCNQ1* gene is causative in the majority of patients. Penetrance is incomplete, which makes genetic testing even more important, since homozygous carriers are at risk of developing arrhythmias when exposed to K^+ channel-blocking drugs. Beta-blockers can prevent arrhythmias. The Valsalva maneuver may cause QT lengthening and tachycardia. Left ganglion stellate block, ablation, or automatic implantable defibrillators may be useful. In the future, treatment with drugs that increase the outward potassium current via other K^+ channels might be possible. Heterozygous carriers of the Jervell and Lange-Nielsen syndrome show slight cardiac dysfunction. The autosomal dominantly inherited long QT syndrome (Romano–Ward syndrome) has the same features – except for the hearing impairment – and is caused by mutations in the same genes.

Mohr Tranebjaerg Syndrome

(also known as: deafness–dystonia–optic atrophy syndrome; Jensen syndrome; DFN1)

MIM   304700

Clinical features   Postlingual progressive SNHI. Renewed clinical investigation of the original families showed Mohr Tranebjaerg syndrome (which was initially named DFN1, and only later changed to Mohr Tranebjaerg) to be a syndromic, rather than a nonsyndromic, disorder. Other features of this progressive syndrome are visual disability (optic atrophy resulting in cortical blindness), ataxia, spasticity, dystonia, fractures (possibly secondary to dystonia), character changes (restlessness, irritability, anxiety, paranoia, and aggressive outbursts), and mental deficiency. The phenotype exhibits variable expression with respect to the presence of the various features (Japanese and American families exhibit no visual loss) and severity of the abnormalities. CT or MRI of the cerebrum may show diffuse central and cortical atrophy. Female carriers can have mild SNHI and minor neuropathy, such as decreased Achilles tendon reflexes, and mild reduction of pain and temperature sensation. In addition, focal dystonias, such as head shaking, writer's cramp, and torticollis, have been reported in female carriers.

Age of onset   SNHI is detected as early as birth up to about 5 years. Visual symptoms begin in childhood; the onset of neurologic symptoms varies from 8 years to early adulthood.

Epidemiology   Approximately 5–10 families have been described.

Inheritance   X-linked recessive

Chromosomal location   Xq22

Genes   *TIMM8A* (translocase of inner mitochondrial membrane 8A; formerly called *DDP* [deafness/dystonia peptide])

Mutational spectrum   Small deletions, missense mutations, a nonsense mutation, and deletion of the entire gene have been described.

Effect of mutation   The missense mutation leads to instability of the protein. All other mutations can be predicted to cause truncation or absence of the protein. This results in defective import of TIMM23, which is important for protein import to the mitochondrial matrix. Therefore, TIMM8A deficiency could indirectly lead to a decrease in oxidative phosphorylation and energy production. Mitochondrial dysfunction apparently leads to neurodegenerative disease, resulting in optic atrophy, loss of neurons in the basal ganglia, abnormal mitochondria in muscle, and a reduced number of spiral ganglion cells in the cochlea.

Diagnosis   In males with SNHI associated with neurologic and/or visual disability, mutation screening in the *TIMM8A* gene is an option, even if the case is sporadic.

Counseling issues   Neurologic and ophthalmologic evaluation should be performed. Audiologic and vestibular tests should be done. In cases where hearing aids cease to provide benefit, cochlear implants have been reported to be successful.

Neurofibromatosis

To date, nine distinct types of neurofibromatosis (NF) have been identified, among which NF1 (von Recklinghausen syndrome) is the most prevalent type, accounting for about 90% of cases. NF1 is defined as six or more café-au-lait spots, cutaneous neurofibromas, iris Lisch nodules (iris hamartomas), and axillary freckling. Acoustic schwannoma may occur unilaterally in about 2%–4% of patients. The disease is inherited in an autosomal dominant fashion, although about 50% of cases represent a new mutation. The defect is located in a tumor suppressor gene.

Within the scope of this book, the phenotype of NF2 (5%–10% of cases) is the most important, as this phenotype frequently exhibits bilateral acoustic schwannomas. However, bilateral acoustic schwannomas have also been reported in some cases of NF3. NF3 is a mixed type that has characteristics of both NF1 and NF2, with a multiplicity of brain tumors. These tumors usually have an early onset with rapid progression. All reported cases have been sporadic.

Neurofibromatosis, Type II

(also known as: NF2; central neurofibromatosis; bilateral acoustic schwannomas; bilateral acoustic neurofibromatosis)

MIM   101000

Clinical features   Bilateral acoustic schwannomas are found in about 95% of cases. The symptoms that arise from this benign, slow-growing tumor are a result of pressure on the vestibulo-cochlear and facial nerves. The tumor is located at the vestibular branch and growth is often unpredictable (see **Figure 17**). SNHI is the presenting symptom in about 50% of cases, and is mostly asymmetrically progressive; however, sudden deafness has been reported. Balance problems, vertigo, tinnitus, facial nerve paresis, and headaches may also develop. Other presenting symptoms are diplopia, dysphonia, and trigeminal nerve hypoaesthesia. Symptoms

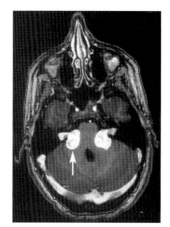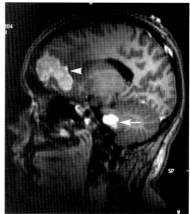

Figure 17. Saggital and axial T₁-weighted MRI after intravenous gadolinium of an 18-year-old woman suffering from neurofibromatosis type II. A large frontal lesion on the right (arrowhead) and a bilateral lesion located in the cerebellopontine angle (arrow indicates right lesion) are visible.

develop unilaterally, and involvement of the other side usually occurs within 2 years. Café au lait spots (40% of patients) and cutaneous tumors (20%–70%) may be present. Other nervous system tumors are found in about 50% of cases, causing symptoms that depend on the location of the tumor. A common finding is benign tumors of the central nervous system (CNS) – mostly schwannomas, neurofibromas, meningiomas, or gliomas. Schwann cell tumors and neurofibromas can also be found along the spinal roots, cranial nerves, and even along peripheral nerves. Raised intracranial pressure may be found, causing chronic papilledema and secondary visual loss. Visual loss can also be caused by damaged optic pathways, retinal hamartomas, and corneal opacities. Juvenile posterior subcapsular lenticular opacities and capsular cortical cataracts are found in 50%–80% of patients.

Age of onset   The onset of SNHI is usually in the second or third decade; however, late onset (up to 75 years) has been reported sporadically.

Epidemiology   The estimated birth prevalence is about 1:33,000–44,000. Symptomatic prevalence is around 1:210,000.

Inheritance   Autosomal dominant

Chromosomal location	22q12.2

Genes   NF2 (neurofibromatosis type II, a tumor suppressor gene)

Mutational spectrum   All types of mutations are found: nonsense, missense, and splice-site mutations, insertions, and deletions. Affected individuals are heterozygous for the mutation in one copy of the tumor suppressor gene. An additional somatic mutation in the other copy is needed to facilitate tumor growth (Knudson's two-hit hypothesis).

Effect of mutation   The majority of mutations result in truncation of merlin (also known as schwannomin), the protein encoded by the NF2 gene. Missense mutations are predicted or shown to change its interaction with other proteins. Merlin acts as a suppressor of cell growth and plays a role in the correct translation of growth signals from other cells or the cell matrix. Merlin also plays a role in cell adhesion, and mutated merlin leads to abnormal organization of the actin cytoskeleton.

Diagnosis   The presence of the clinical features will indicate the diagnosis. Tone and speech audiograms, brainstem-evoked responses (BER), and an electronystagmogram (ENG) are usually performed. Neuroimaging (brain and spine) may reveal tumors in the CNS. Mutation analysis can lead to the diagnosis – this test is positive in about 84% of familial cases and 50% of sporadic cases.

Individuals with the following clinical features are clinically considered to have NF2:

• bilateral vestibular schwannomas

or

• family history of NF2 (first-degree family relative); plus

  – unilateral vestibular schwannoma <30 years; or

  – any two of the following: neurofibroma, meningioma, glioma, schwannoma, juvenile posterior subcapsular lenticular opacities/juvenile cortical cataract

Individuals with the following clinical features are advised to be evaluated for NF2 (presumptive or probable NF2):

• unilateral vestibular schwannoma <30 years; plus

- at least one of the following: meningioma, glioma, neurofibroma, schwannoma, juvenile posterior subcapsular lenticular opacities/juvenile cortical cataract

or

- multiple meningiomas (two or more); plus
  - unilateral vestibular neuroma <30 years; or
  - one of the following: glioma, neurofibroma, schwannoma, juvenile posterior subcapsular lenticular opacities/juvenile cortical cataract

Bilateral facial nerve schwannomas in the absence of an affected first-degree relative may also be caused by the *NF2* gene.

Counseling issues

NF2 is fully penetrant by the age of 60 years. Intrafamilial phenotypic variability is usually low. About 50% of cases are sporadic. Inactivation of the *NF2* gene is also found in almost all sporadic schwannomas. Clinical assessment may consist of cutaneous, otorhinolaryngologic, neurologic, and ophthalmologic examinations, and full CNS imaging. In cases where a "wait and see" policy is followed, neurologic examination, MRI scans, audiograms, and sometimes BER can be obtained on a regular basis. Most patients become completely deaf. Note that vestibular schwannomas in these patients grow more often and more rapidly than sporadic unilateral schwannomas. Since NF2 vestibular schwannomas are more likely to infiltrate than other schwannomas, it is more difficult to preserve facial and cochlear nerve function after surgery. Other disabling features are poor balance, visual problems, and weakness due to spinal tumors; the latter causes many patients to be wheelchair-bound. It is important to realize that therapy in these patients is focused mainly on palliative care as a result of the multifocality of the disease. Microsurgical resection and stereotactic radiosurgery are the treatment modalities available. Each patient requires an individualized therapeutic approach. Cochlear implantation or brainstem implantation (in cases where the cochlear turns are inaccessible or the spiral ganglion or cochlear nerve is severely injured) are options to rehabilitate a deaf patient.

In sporadic cases, it is impossible to predict the exact chance that offspring will be affected. In sporadic cases, the disease can be generalized or localized (ie, disease in one part of the CNS). The gonads are involved in a number of cases, and this may result in affected offspring. Patients with the sporadic localized or even the generalized phenotype are thought to develop less severe disease than that found in patients with the inherited type.

Noonan Syndrome, Type I

(also known as: NS1; Turner-like syndrome; male Turner syndrome)

MIM 163950

Clinical features Webbed neck, pectus excavatum, cryptorchidism, and pulmonic stenosis (see **Figure 18**). The neck is webbed and the posterior hairline low in 55% of patients. Seventy percent of patients have a pectus excavatum or carinatum. Short stature with prenatal onset is found in about 50%.

Figure 18. Affected father and son. The father does not show the typical facial features. However, his son shows a webbed neck, low nasal bridge, hypertelorism, and epicanthal folds. The auricles are posteriorly angulated and low set.

The mouth often has an abnormal width, the upper lip protrudes, and retrognathia and malocclusion are found. As the patient ages, the shape of the face becomes triangular, thereby masking the initial anomalies. The nasal bridge might be low. Occasionally, a high-arched palate or cleft palate and an asymmetric head are found.

Ocular symptoms consist of epicanthal folds, hypertelorism (95% of patients), ptosis, myopia, downslanting palpebral fissures, strabismus, keratoconus, and nystagmus.

The auricles may be abnormally shaped (90%), posteriorly angulated, and low set. Profound SNHI is occasionally found, as are congenital anomalies of the ossicular chain, such as absence of the long process of the incus.

Mental retardation has been reported in approximately 25% of cases. The heart shows pulmonary valve stenosis and/or other anomalies (about 65%). Cryptorchidism (probably causing infertility when bilaterally present) is often accompanied by a micropenis. Bleeding disorder is sometimes present (20%).

Age of onset	Congenital
Epidemiology	1:1,000–2,500 births
Inheritance	Autosomal dominant
Chromosomal location	12q24.1
Genes	*PTPN11* (protein tyrosine phosphatase, nonreceptor-type 11; also known as *SHP2*)
Mutational spectrum	All mutations are amino-acid substitutions, most of which are present in the region of the gene encoding the Src homology region 2 (SH2) or phosphatase domain of the protein.
Effect of mutation	The mutations are predicted to shift the balance between the active and inactive conformations of the protein towards the active state.
Diagnosis	Clinical features may point to a diagnosis and mutation analysis in the *PTPN11* gene can be performed. It has been suggested that metacarpophalangeal pattern profile (MCPP) analysis can lead to a correct classification in approximately 93% of patients.
Counseling issues	About 50% of NS cases are sporadic; however, autosomal dominant inheritance is also frequently reported. Large families with NS are rare. A marked variability in the phenotype has been reported. NS is genetically heterogeneous. An autosomal recessive form of NS, in which there is a higher prevalence of hypertrophic obstructive cardiomyopathy, has been postulated. The final height of children with NS does not appear to be substantially increased by the use of growth hormone. Laboratory findings include deficiency of various factors (factors VIII, XI, XII) and thrombocytopenia, causing bleeding disorders. Feeding difficulties may necessitate nasogastric tube feeding in many children.

Norrie's Disease

MIM	310600
Clinical features	Specific ocular symptoms, progressive hearing impairment, and mental retardation. Progressive ocular symptoms involve congenital, usually bilateral pseudoglioma of the retina, nonattachment of the retina, retinal hyperplasia, hypoplasia and necrosis of the inner layer of the retina, cataracts, blindness, iris atrophy, iris synechiae, and phthisis bulbi. Less than 50% of patients are mentally retarded or hearing impaired. Psychosis, increased susceptibility to infections, microcephaly, cryptorchidism, and hypogonadism may also be found.
Age of onset	Congenital progressive. The ocular symptoms are usually observed first. SNHI may develop in the second decade.
Epidemiology	About 300 cases have been described to date.
Inheritance	X-linked recessive
Chromosomal location	Xp11.4
Genes	*ND* (Norrie's disease; also known as norrin)
Mutational spectrum	Deletions of the entire gene, intragenic deletions, insertions, and nonsense and missense mutations have been detected.
Effect of mutation	The encoded protein, norrin, seems to bind very strongly to the extracellular matrix. Mutations lead to absence of the norrin protein or a protein with no or aberrant function. Several lines of evidence support the hypothesis that norrin functions in developmental vasculogenesis of the retina. The protein is thought to be secreted from the cell and to interact with the extracellular matrix.
Diagnosis	Diagnosis is often difficult because ocular symptoms are the only feature in >50% of patients. Mutation analysis will lead to a final diagnosis.
Counseling issues	Penetrance is probably full. Only males are affected; females who carry the mutation are unaffected. Fathers are unable to pass the disease onto their sons. There is large intra- and interfamilial phenotypic variability, which makes it difficult to determine a genotype–phenotype correlation. Mutations at the C-terminal site of the gene result in a milder phenotype.

Oculo-auriculo-vertebral Spectrum

(also known as: hemifacial microsomia; Goldenhar syndrome; Goldenhar–Gorlin syndrome; first arch syndrome; first and second branchial arch syndrome; lateral facial dysplasia; oral–mandibular–auricular syndrome; otocranialcephalic syndrome; auriculobranchiogenic dysplasia; craniofacial microsomia; facio-auriculo-vertebral dysplasia; otomandibular dysostosis; necrotic facial dysplasia; intrauterine facial necrosis; hemignathia and microtia syndrome; unilateral facial agenesis)

MIM	164210
Clinical features	Mainly unilateral, affecting aural, oral, and mandibular (100%) development (first and second branchial arch). There is large phenotypic variation, ranging from mild to severe. It may present bilaterally (10%–33%), with one side (most often the right) more affected than the other. Major facial asymmetry is present in 20% of cases; however, milder asymmetry is found in approximately 65%. The main cause of the asymmetry is unilateral abnormality (see **Figure 19**) and/or displacement of the pinna and asymmetry of the facial skeleton. In young children, the latter becomes apparent by the age of about 4 years. Maxillary, temporal, malar, and mandibular bones may show hypoplasia. Isolated microtia, and auricular or preauricular abnormalities often represent the mildest manifestation. About 50% of cases have been reported to show other anomalies in addition to the main characteristics. The syndrome may be accompanied by vertebral anomalies and epibulbar dermoids, in which case it is called "Goldenhar syndrome" (see **Figures 20** and **21**).

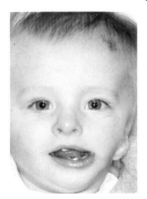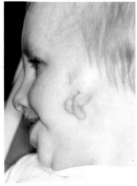

Figure 19. Severe unilateral expression of the oculo-auriculo-vertebral spectrum. Only a remnant of the left auricle is present. Figure courtesy of R Admiraal.

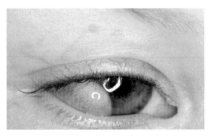

Figure 20. A typical feature of Goldenhar syndrome is the epibulbar dermoid.

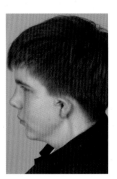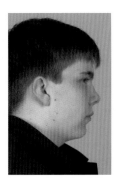

Figure 21. A patient exhibiting Goldenhar syndrome on the left side. Hypoplasia of the facial skeleton and auricle are obvious. The auricle is also low-set and the meatus shows partial atresia. Besides this, the patient suffers from left-sided conductive hearing impairment and multiple abnormalities of the spine.

Eye symptoms consist of narrowing of the palpebral fissure, palpebral shortening, coloboma of the upper eyelid, cataract, and sporadically anophthalmia, microphthalmia, and retinal abnormalities. Epibulbar tumors are found in about 35% of patients, and vary in size up to 8–10 mm.

Auricular abnormalities are found in around 65% of cases and vary from mild dysmorphology to anotia. Preauricular tags of skin and cartilage are very common, as are preauricular sinuses.

Conductive impairment and SNHI have been reported in over 50% of patients, with SNHI only being found in 15% of cases. Abnormalities are found in the middle ear, external ear, ossicles, and skull base. A narrowed external meatus is found in the mild type, with the severe type showing complete atresia. The ears may be low set.

Among the various oral manifestations, cleft lip and/or palate (7%–15%), velopharyngeal insufficiency (35%), macrostomia (40%), and malocclusion have been reported.

A wide spectrum of central nervous system malformations may be found, including severe anatomical abnormalities, cranial nerve dysfunction (10% have facial nerve weakness), and mental deficiency (5%–15%). Cardiac, skeletal, lung, renal, and other abnormalities can also occur.

The phenotype has been found in children of diabetic mothers. Infants are often relatively small and may have feeding difficulties related to their cleft lip/palate or narrowed pharyngeal airway. Obstructive sleep apnea has also been reported.

Age of onset	Congenital
Epidemiology	Probably about 1:5,600 births. The male to female ratio is approximately 3:2. The ratio of affectation of the right versus the left ear is also 3:2. This syndrome represents the second most common facial anomaly syndrome after cleft lip and/or cleft palate.
Inheritance	Usually sporadic. Familial cases that show obvious intrafamilial variation have been reported. Autosomal dominant and autosomal recessive inheritance have been reported, the former accounting for 1%–2% of cases.
Chromosomal location	Unknown, but several chromosomal aberrations have been described in association with the syndrome/spectrum. There are linkage data suggestive for a locus in 14q32.
Genes	Unknown
Mutational spectrum	Unknown
Effect of mutation	The primary defect is unknown, but the phenotype may be caused by locoregional (first and second branchial arch) disruption of the vascular supply at about 30–40 days gestation. Another theory suggests that disruption of neural crest cell migration and distribution is involved.
Diagnosis	Diagnosis can usually be made by evaluation of the clinical features, sometimes supported by chromosomal analysis (eg, for trisomy 22).

Counseling issues	The disease is very heterogeneous in cause and expression pattern. The overall recurrence risk is about 2%–3%; however, recurrence risk counseling should be given on an individual basis. First-degree relatives should be evaluated, keeping in mind that expression can be mild. CT scanning of the temporal bone is essential as part of the hearing evaluation. Because of the wide range of abnormalities, evaluation should be multidisciplinary. Patients with features of the oculo-auriculo-vertebral spectrum and features overlapping with Townes–Brocks syndrome (dysplastic ears, ear tags, hearing impairment, thumb anomalies, anal defects, and renal anomalies; MIM 107480) should be tested for *SALL1* mutations, which cause Townes–Brocks syndrome.
	Surgical procedures are of major importance to correct life-threatening abnormalities. Cosmetic and functional surgical management can involve excision of preauricular tags, transposition of the microtic ear, surgical reconstruction of the ear, and closure of macrostomia. Surgery of the mandible includes elongation of the ramus ascendens or reconstruction of a missing ramus, including condyle reconstruction. An orthodontic splint can be used to regulate and stimulate growth of the teeth and maxilla. After skeletal framework reconstruction, any remaining soft-tissue asymmetry can be corrected by bone grafts or free-tissue transfer. Ramus elongation can also be achieved by mechanical mandibular dystraction after performing mandibular osteotomy and placing metallic pins in the mandible. Conductive hearing impairment can often be corrected surgically; a bone-anchored hearing aid is needed in only a small number of patients.

Osteogenesis Imperfecta

(a) Osteogenesis Imperfecta, Type I

(also known as: osteogenesis imperfecta tarda; Lobstein's disease;
osteogenesis imperfecta with blue sclerae)

MIM	166200
Clinical features	Mild to moderately severe bone fragility, blue sclerae, hyperextensibility, and hearing impairment. Growth deficiency is found in about 50% of cases, usually with postnatal onset; in adulthood, 50% of patients have a height less than the third percentile. Skin and sclerae are thin, and

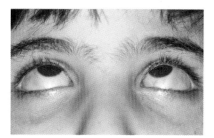

Figure 22. A patient with osteogenesis imperfecta type I gazing upwards. Blue sclerae are a typical feature of this syndrome.

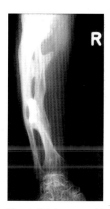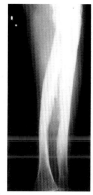

Figure 23. Increased bone fragility is a feature of all four types of osteogenesis imperfecta. Multiple fractures can lead to bowing and progressive deformity of bones.

visualization of the choroid results in blue sclerae (100% of patients) (see **Figure 22**). The skin and eyes are easily bruised in about 75% of patients. Hypoplasia of dentin and pulp and translucency of the teeth, resulting in amber or blue–gray coloration, may be found in types IB and IC; however, dentinogenesis imperfecta is uncommon in type I (although it is common in type IV). Types IB and IC are differentiated on the basis of specific dental abnormalities – caries are often found, and teeth show late eruption and are irregularly placed. Type IA shows the same phenotype without dental abnormalities. Macrocephaly (18%), wormian bones, maxillary hypoplasia (relative mandibular prognathism), temporal bulging, and a triangular face (30%) can also be found.

There may be mild limb deformity (20%) with postnatal onset, showing anterior or lateral bowing of femora and anterior bowing of the tibia (see

Figure 23). Fractures appear in about 92% of patients, and scoliosis and kyphosis both in about 20%. Multiple fractures are found in about 90%; however, patients with type IB appear to have a higher fracture rate and are more likely to have a height below the second percentile than type IA. Fractures are first noted at birth in 8% of cases, in the first year in 23%, before the age of 6 years in 45%, and during school years (6–18 years) in about 17%. The prevalence of fractures diminishes after adolescence; however, the frequency may increase in postmenopausal women and men aged >60 years. Loss of stature may be noted in adults. Joint hyperextensibility is found in all patients.

Hearing impairment due to otosclerosis is found in around 50% of patients <30 years, rising to 95% in patients aged >30 years. Onset of the conductive component is usually in the late second and third decades of life. SNHI is found in approximately 50% of all patients. Mixed hearing impairment, and especially SNHI, can be progressive. Progression has been estimated to be about 1 dB/year, leading to profound hearing impairment. The conductive component is caused by immobility of the stapes footplate or by a fracture of the crura or atrophy of the stapes. Fracture of the malleus handle has also been reported. The sensorineural component may, in addition, be caused by ossification in the otic capsule or hemorrhage in the inner ear. Tinnitus and vertigo are other common features. Syndactyly, aortic root dilatation (without dissection), or a floppy mitral valve (18%) are occasionally found.

Age of onset	Onset is in early childhood for most features.
Epidemiology	1:10,000–25,000. However, the mild presentation in many patients suggests that the disease is underdiagnosed.
Inheritance	Autosomal dominant
Chromosomal location	17q21.31–q22.1, 7q22.1
Genes	*COL1A1* (type I collagen α1 chain), *COL1A2*; other genes might also be involved.
Mutational spectrum	Substitutions for glycine are found in both genes. In addition, splice-site mutations and other mutations leading to a null allele of the *COL1A1* gene have been described.

Effect of mutation	For *COL1A1*, most nonsense mutations, insertions, and deletions lead to a reduction in the amount of type I collagen (haploinsufficiency). The effects of splice-site mutations and of substitutions of glycine residues are difficult to predict.
Diagnosis	Recognition of the clinical characteristics may lead to the diagnosis. In most cases, mutation analysis is helpful. Reduced synthesis of procollagen 1 by dermal fibroblasts can also reveal the diagnosis. Ultrasound, chorionic villus sampling, and amniocentesis can be used to reach a prenatal diagnosis.
Counseling issues	There is large intra- and interfamilial variability in expression. A paternal age effect has been suggested for new mutations. Hearing impairment may be treated by the fitting of hearing aids or with surgery (stapedotomy). No medical treatment exists to increase the amount of procollagen 1. It has been suggested that prolonged intravenous infusion of pamidronate (together with elemental calcium and vitamin D supplementation) might reduce the number of fractures and increase bone density. Fractures can be treated with standard orthopedic procedures, which usually result in normal healing. Management of fracture-associated pain is of major importance. Postmenopausal women may be given physiotherapy and hormone replacement therapy.

(b) Osteogenesis Imperfecta, Type IIA

(also known as: osteogenesis imperfecta congenita; Vrolik type of osteogenesis imperfecta)

MIM	166210
	Osteogenesis imperfecta types IIA–C, caused by mutations in either the *COL1A1* (type I collagen α1 chain) or *COL1A2* genes, cause severe bone fragility and blue sclerae. Children are often stillborn or die shortly after birth. Approximately 90% of children die before the age of 4 weeks. The disease is due to new dominant mutations, resulting in dysfunction of the protein.

(c) Osteogenesis Imperfecta, Type III

MIM	259420
Clinical features	Moderately severe to severe bone fragility and normal sclerae (blue sclerae in infancy). Fractures are found at birth in >50% of patients; by the age of 2 years, all patients have numerous fractures. The disease is progressively deforming. Death may occur in the first decades of life, usually as a result of progressive kyphoscoliosis causing cardiopulmonary insufficiency. Head size is proportionally large, and frontal and temporal bossing lead to triangular facies. Blue sclerae are present in infancy, but fade with increasing age. Vertebral flattening, trunk shortening, and severe kyphoscoliosis may be found. Short stature is present at birth, and adult height varies between 92 and 108 cm. Joint hyperextensibility is frequently found. Most patients are handicapped, and without aggressive intervention will become wheelchair-bound. Hearing impairment is found in only about 5% of cases and is of conductive origin. Audiologic findings have not been very well documented.
Age of onset	Congenital
Epidemiology	Rare. The autosomal recessive type is the most common form of osteogenesis imperfecta in South African blacks. Studies have shown that, in Australia, the ratio of type I to type III osteogenesis imperfecta is about 7:1. In contrast, this was found to be about 1:6 in South African blacks.
Inheritance	Autosomal dominant, in some cases autosomal recessive
Chromosomal location	17q21.31–q22.1, 7q22.1
Genes	*COL1A1* (type I collagen α1 chain), *COL1A2*
Mutational spectrum	Substitutions for glycine and exon-skipping.
Effect of mutation	Mutations lead to dysfunctional type I collagen and aberrant collagen formation. The recessive mutations in *COL1A2* prevent chain association (ie, the ability of α2 chains to be incorporated into type I procollagen is altered).

Diagnosis	Ultrasound, chorionic villus sampling, and amniocentesis can be used for prenatal diagnosis.
Counseling issues	Prolonged intravenous infusion of pamidronate has been shown to increase bone density. Whether this leads to a reduction in fracture fragments is under investigation. Fractures can be treated with standard orthopedic procedures, which usually result in normal healing. Intramedullary rods may diminish bowing and angulation deformities. Management of fracture-associated pain is of major importance.

(d) Osteogenesis Imperfecta, Type IV

MIM	166220
Clinical features	Mild to moderately severe bone fragility, normal sclerae (pale-blue sclerae may be detected in early childhood), hearing impairment, and often opalescent teeth. The severity of this type is in between that of types I and III.
	Fractures are found at birth in about 25% of patients. Most fractures occur during childhood and the frequency diminishes after puberty. Some patients never have fractures. Stature is often below the 5th percentile. Maxillary hypoplasia (relative mandibular prognathism), temporal bulging, and a triangular face may be found, along with wormian bones, kyphoscoliosis, and flattened vertebrae. Joint hyperextensibility is frequent. Some patients bruise easily. Hearing impairment occurs in about 30% of patients >30 years of age; this can be primarily conductive, primarily sensorineural, or mixed. Type IVA patients have normal teeth, while those with the more common type IVB have opalescent teeth.
Age of onset	Congenital
Epidemiology	Unknown
Inheritance	Autosomal dominant
Chromosomal location	17q21.31–q22.1
Genes	*COL1A1* (type I collagen α1 chain), *COL1A2* (more common)

Mutational spectrum	Missense mutations (substitutions for glycine), splice-site mutations leading to exon skipping, and partial gene deletions are found.
Effect of mutation	Mutations lead to abnormal functioning of type I collagen and abnormal collagen fibrils.
Diagnosis	The clinical features can lead to the diagnosis, which can be confirmed by mutation analysis. Ultrasound, chorionic villus sampling, and amniocentesis can be used to reach a prenatal diagnosis.
Counseling issues	The intrafamilial variability of the phenotype can be striking. A paternal age effect has been suggested for new mutations. Hearing impairment due to otosclerosis may be treated with a hearing aid or surgery (stapedotomy). Prolonged intravenous infusion of pamidronate has been shown to increase bone density. Whether this leads to a reduction in fracture fragments is under investigation. Fractures may be treated with standard orthopedic procedures, which usually result in normal healing. Management of fracture-associated pain is of major importance. Postmenopausal women may be given physiotherapy and hormone replacement therapy.

Osteopetrosis

(a) Osteopetrosis, Autosomal Recessive

(also known as: severe autosomal recessive osteopetrosis; autosomal recessive Albers–Schönberg disease; autosomal recessive marble bones)

MIM	259700
Clinical features	Increased osseous density and compression of cranial foramina. Nearly all bones are affected. Bone marrow compression leads to anemia and hepatosplenomegaly (due to extramedullary hematopoiesis; 50% of patients). Phagocytic dysfunction can result in severe infections. Compression of cranial foramina can lead to blindness (pressure on the optic veins leads to optic atrophy), hearing impairment, facial paralysis, vestibular nerve dysfunction, and extraocular muscle paralysis. Osteomyelitis might also be found. The affected bones are expanded and dense. In addition, the skull is thickened and dense. There may be macrocephaly and hydrocephalus. The facial bones also appear to be

more dense than usual, giving the appearance of a square head. The calvariae and paranasal sinuses are poorly aerated. Fractures are commonly found. Oral pathology includes distortion of the permanent teeth, which often fail to erupt, and osteomyelitis after dental extraction (due to an insufficient blood supply). Mental retardation is found in about 20% of cases. Hypocalcemia may result in seizures. About 25%–50% of patients show moderate mixed hearing impairment with onset in childhood. A narrowed epitympanic space with epitympanic fixation of the incudomalleal joint is often seen. Children are often stillborn or die within several months of birth. Only 30% of patients survive to the age of 6 years. No patient has survived adolescence.

Age of onset	Congenital
Epidemiology	–
Inheritance	Autosomal recessive
Chromosomal location	11q13.4–q13.5, 16p13.3, 6q21
Genes	*TCIRG1* (T-cell immune regulator 1), *CLCN7* (chloride channel 7), *GL* (gray-lethal)
Mutational spectrum	Small deletions, and nonsense and splice-site mutations have been found in *TCIRG1*. All mutations can be predicted to lead to protein termination and thus are considered to be loss of function mutations. A missense and nonsense mutation have been described in *CLCN7*. These are also considered to be loss of function mutations. A deletion has been found in *GL*, resulting in complete loss of function
Effect of mutation	Osteoclasts fail to resorb primary spongiosa, resulting in an increased osseous density. The TCIRG1 protein functions as a proton pump and the CLCN7 protein as a chloride channel. GL seems to play an important role in osteoclast and melanocyte maturation and function.
Diagnosis	The clinical features may lead to a diagnosis. X-ray evaluation can lead to a prenatal diagnosis.
Counseling issues	Allogenic bone marrow transplantation has been reported to be successful in some cases. Interferon γ-1b can also be administered in

patients who cannot be transplanted or who are awaiting transplantation. This therapy results in increased bone marrow space and hemoglobin, and a decrease in the infection rate. Conductive hearing impairment may be corrected by surgery of the ossicular chain. A mild form of autosomal recessive osteopetrosis (MIM 259710) has been reported in a few cases. The phenotype of the mild form resembles that of the autosomal dominant form of osteopetrosis (see p. 99). The corresponding genetic locus is unknown.

(b) Osteopetrosis with Renal Tubular Acidosis

(also known as: Guibaud–Vainsel syndrome)

MIM	259730
Clinical features	Fractures, short stature, mental retardation (sometimes), dental malocclusion, hepatosplenomegaly, and optic nerve compression with loss of vision. Initial mild anemia and osteopetrosis improve to a certain degree at puberty. Mild tubular acidosis is found. During adolescence, the basal ganglia develop calcifications. Periodic hypokalemic paresis can be another feature. This type seems to be less severe than osteopetrosis without renal tubular acidosis (MIM 259700).
Age of onset	This disorder manifests within the first 2 years with fractures.
Epidemiology	This is a rare disease, of which more than half of the known cases have been observed in families from Kuwait, Saudi Arabia, and North Africa.
Inheritance	Autosomal recessive
Chromosomal location	8q22
Genes	*CA2* (carbonic anhydrase II)
Mutational spectrum	Missense, nonsense, and splice-site mutations, and small insertions and deletions are found.
Effect of mutation	The detected mutations lead to either amino-acid substitutions or premature protein termination. For a number of mutations, carbonic anhydrase II enzyme activity was tested and shown to be strongly reduced. Therefore, acid–base homeostasis is disturbed in the affected

tissues. In bone, this leads to a failure of bone resorption. In the kidney, this leads to acidosis of the proximal and distal tubule.

Diagnosis   The clinical features and mutation analysis lead to the diagnosis.

Counseling issues   The diagnosis cannot be made radiologically for a fetus because osteopetrosis does not appear prenatally. At birth, carbonic anhydrase levels are extremely low in the erythrocytes of healthy children. Therefore, diagnosis is not possible by measuring enzyme activity before or soon after birth.

(c) Osteopetrosis, Autosomal Dominant, Types I and II

(also known as: osteosclerosis fragilis generalisata; autosomal dominant Albers–Schönberg disease; autosomal dominant marble bones)

MIM	166600
Clinical features	Backache, headache, and, sporadically, trigeminal neuralgia. Anemia, hepatosplenomegaly, blindness, and mental retardation are not found. Cranial nerve (II, III, VII) palsies are found in about 10%–20% of patients. In type II, fractures are found in 60%–80% of patients. The skull base is dense, and sclerosis is also found in the spine and pelvis in type II patients. Type I shows thick and sclerotic calvariae. In type I, fractures are rarely found and fifth cranial nerve palsy is almost only found in this type; seventh cranial nerve palsy is more frequently found in type II. Most patients with type II show conductive hearing impairment; in type I, about 50% show such an impairment. This disease is asymptomatic in 20%–40% of patients.
Age of onset	Later in life than the recessive types, but within the first few years of life.
Epidemiology	Unknown, but more common than the recessive types.
Inheritance	Autosomal dominant
Chromosomal location	Type I: 11q13.4 Type II: 16p13.3
Genes	Type I: *LRP5* (lipoprotein receptor-related protein 5) Type II: *CLCN7* (chloride channel 7)

Mutational spectrum	Type I: Two missense mutations involving glycine 171 have been found.
	Type II: Missense mutations have mainly been described, along with two small deletions.
Effect of mutation	Type I: Mutations in the *LRP5* gene lead to disturbance of the Wnt signaling pathway. Wnt-mediated signaling via *LRP5* affects bone accrual during growth.
	Type II: The mutations in *CLCN7* are thought to have a dominant-negative effect and lead to impaired acidification of the extracellular resorption lacuna.
Diagnosis	Clinical characteristics and X-rays may lead to the diagnosis. Mutation analysis or linkage studies can confirm this.
Counseling issues	There is large intra- and interfamilial variability.

(d) Differential Diagnoses to Osteopetrosis

A bone disease similar to osteopetrosis is hyperostosis corticalis generalisata (van Buchem disease, MIM 239100). This is inherited in a recessive way, linked to chromosome 17q11.2. In van Buchem disease, apposition of cortical bone with a normal structure leads to bone thickening. Cortical hyperostosis of the skull, mandible, clavicle, and ribs are characteristic. Compression of the cranial foramina can lead to progressive loss of vision, hearing impairment, and facial palsy. Mixed hearing impairment is usually progressive. Compression of the eighth nerve results in a perceptive component; thickening and fixation of the ossicles accounts for the conductive component. Van Buchem's disease should be differentiated from sclerosteosis (MIM 269500), which shows gigantism and hand abnormalities, in addition to the above features. In sclerosteosis, *SOST* (sclerostin) on chromosome 17q12–q21 is the disease-causing gene, and the disease is recessively inherited.

Osteitis deformans (morbus Paget, MIM 602080) is a metabolic disease characterized by excessive bone resorption and formation due to activated osteoclasts. The disease is dominantly inherited. Various loci are involved, located on chromosomes 18q22.1, 5q35–qter, 5q31, and 6p21.3. For the locus 18q22.1, *TNFRSF11A* (tumor necrosis factor receptor superfamily, member 11A) has been identified as the disease-causing gene. The age of onset is usually >40 years. The disease is characterized by bone pain, fractures, hearing impairment (up to 40% conductive impairment and/or SNHI), and neurologic complaints. Bone deformities and enlargement are found at the pagetic sites (mostly axial skeleton). Environmental factors and virus infection might also be involved in the etiology.

Pendred Syndrome

MIM	274600
Clinical features	Moderate to profound prelingual SNHI and (usually) a variable goiter with a positive perchlorate test. Most patients are euthyroid, while a minority are hypothyroid. Thyroid-stimulating hormone (TSH) levels are often in the upper end of the normal range; serum thyroglobulin is often significantly elevated at a young age (note that the extent of the elevation is very variable). Subtotal thyroidectomy because of tracheal compression often results in recurrence of goiter. Up to one third of affected adults never manifest clinical signs of goiter. In many cases, SNHI is progressive and fluctuating, and a vestibular dysfunction can be found. Other characteristics include hypoplasia of the cochlea (Mondini malformation), an enlarged vestibular aqueduct (>80% of patients), and an enlarged endolymphatic sac (see **Figure 24**). Specific mutations in the gene responsible for this disorder may also cause the recessively inherited nonsyndromic DFNB4 hearing impairment and enlarged vestibular aqueduct (EVA) syndrome.
Age of onset	SNHI is congenital. Goiter can occur at any age, but usually develops after puberty.
Epidemiology	The prevalence is about 0.8–7.5:100,000 people, ie, 0.8%–7.5% of all early childhood SNHI. Therefore, Pendred syndrome is probably the most prevalent form of syndromic SNHI. Prevalence might be underestimated because of undiagnosed Pendred syndrome in patients with a mild, nonclassic phenotype.
Inheritance	Autosomal recessive, variable expression
Chromosomal location	7q31–q34

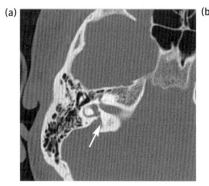 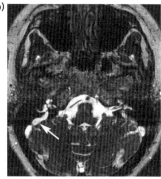

Figure 24. (a) Axial CT scan of the right ear of a patient showing an enlarged vestibular aqueduct (arrow). (b) MRI scan of the same ear revealing an enlarged endolymphatic sac (arrow) to the medial side, leading to an enlarged endolymphatic duct that ends in the sacculus of the labyrinth.

Genes	*SLC26A4* (solute carrier family 26, member 4; formerly called *PDS* [pendrin])
Mutational spectrum	Missense and splice-site mutations, insertions, and deletions are found in the *SLC26A4* gene.
Effect of mutation	The mutations cause an aberrant protein (SLC26A4 protein) either by the substitution of an amino acid or by disruption of the protein, mostly in the C-terminal region. These mutations are predicted to lead to a defect in chloride and iodide anion transmembrane transport, causing impaired thyroxine production due to a defect in organification (localized in the colloid space) of iodine. SLC26A4 protein is localized at the apical membrane of thyrocytes and transports iodide from the thyrocytes into the colloid space. It seems that specific mutations lead to various degrees of ion-transport dysfunction, thereby resulting in either Pendred syndrome (absent function) or EVA syndrome/DFNB4 (reduced function). *SLC26A4* is expressed in the endolymphatic duct and sac, in the utriculus and sacculus, and in the external sulcus in the endolymphatic duct. An abnormal fluid flux in the inner ear, mediated by the defective anion transporter, may be responsible for the enlarged vestibular aqueduct.
Diagnosis	A combination of audiologic, vestibular, and radiologic features and a positive perchlorate test (although the latter is not very sensitive in some cases) are often necessary for diagnosis. Mutation analysis of the

SLC26A4 gene can also be important in obtaining a final diagnosis. Radiologic analysis shows hypoplasia of the cochlea (CT, MRI), a widened vestibular aqueduct (CT, MRI), and an enlarged endolymphatic sac (MRI). These features are the most sensitive diagnosis for this syndrome.

Counseling issues	Diagnosis can be difficult and no single investigation can definitively diagnose Pendred syndrome. Mutation analysis of the *SLC26A4* gene detects about 75% of mutant alleles. In patients of northern European descent, four mutations – L236P, T416P, E384G, and the splice-site mutation 1001+1G>A – account for 72% of the mutations. Patients without the classical combination of features may also have mutations in the *SLC26A4* gene (DFNB4 and EVA syndrome). Families that initially seem to have DFNB4 can later appear to have EVA syndrome or develop goiter, which changes the diagnosis into Pendred syndrome. Hormone replacement therapy is frequently needed to prevent growth of the thyroid into a goiter. See also EVA syndrome (next entry).

Enlarged Vestibular Aqueduct Syndrome

(also known as: EVA syndrome; DFNB4)
Note: it appears that patients classified as having DFNB4 are actually patients suffering from EVA syndrome.

MIM	603545
Clinical features	An enlarged vestibular aqueduct associated with fluctuating, progressive, moderate to severe SNHI and disequilibrium symptoms. SNHI may be progressive to profound. There is major overlap between DFNB4 and EVA syndrome. Future research may reveal that the two diseases represent one and the same entity. One Indian family has been described as having true DFNB4; however, three of those who underwent CT scanning of the temporal bone were shown to have an enlarged vestibular aqueduct. Perchlorate tests were not performed.
Age of onset	Congenital
Epidemiology	Unknown
Inheritance	Autosomal recessive

Chromosomal location	7q31
Genes	SLC26A4 (solute carrier family 26, member 4; formerly called PDS [pendrin])
Mutational spectrum	Missense mutations, an insertion, and a deletion have been found in the SLC26A4 gene.
Effect of mutation	See Pendred syndrome (previous entry).
Diagnosis	CT and/or MRI scans will confirm vestibular aqueduct enlargement. MRI may also identify endolymphatic sac enlargement. Mutation analysis can be performed.
Counseling issues	Disease expression can be variable. SNHI can progress with head trauma, air travel, and diving. Increases in cerebrospinal fluid pressure, along the vestibular aqueduct, may increase the pressure and therefore the local composition of the perilymph, causing progression of SNHI. Close examination should identify features of Pendred syndrome or less prevalent features of branchio-oto-renal (BOR) syndrome (a differential diagnosis; see p. 60). In a group of patients with hearing impairment and an enlarged vestibular aqueduct on CT, about 72% were shown to suffer from Pendred syndrome, while an SLC26A4 mutation (homo- and heterozygosity) was involved in about 86%. In this cohort, BOR syndrome was found in about 3.5%.

Marshall Syndrome

Both Stickler syndrome (next entry) and Marshall syndrome (MIM 154780) are dominantly inherited chondrodysplasias with SNHI, high myopia, and mid-facial hypoplasia (see **Figure 25**). There is an ongoing debate about whether they are distinct entities. Patients with Marshall syndrome are said to show hypertelorism and abnormalities of ectodermal derivates (sparse hair, eyebrows, and eyelashes; diminished ability to sweat). It seems that this syndrome is caused by variable expression of at least two different Stickler genes (COL2A1 and COL11A1), suggesting that it is not a distinct entity. Some patients present with phenotypes of both Stickler and Marshall syndrome.

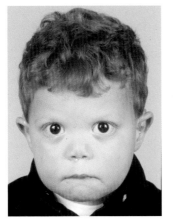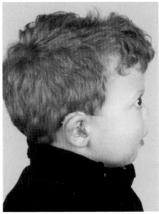

Figure 25. A patient with Marshall syndrome showing midfacial hypoplasia, hypertelorism, severe myopia, and SNHI. A feature of midfacial hypoplasia is the short, upturned nose with a depressed nasal bridge.
Figure courtesy of R Admiraal.

Stickler Syndrome

Three types of Stickler are described below, but a fourth Stickler gene probably exists with a phenotype without severe myopia and retinal detachment.

In patients diagnosed with Pierre Robin syndrome (MIM 261800), the possibility of Stickler syndrome also being present should be kept in mind, especially in familial cases. The features of Pierre Robin syndrome consist of, among others, glossoptosis, micrognathia, cleft palate, upper airway obstruction, and cor pulmonale.

(a) Stickler Syndrome, Type I

MIM   108300

Clinical features   Progressive myopia, vitreoretinal degeneration, cataracts, retinal detachment and blindness, premature joint degeneration (abnormal epiphysis), irregularities of vertebral bodies, midfacial hypoplasia, cleft palate, mitral valve prolapse (up to 50%), and high-frequency, often mild, nonprogressive SNHI. This SNHI is found in about 50% of patients and is usually less severe than in the Stickler syndrome type II phenotype. Facial involvement consists of a flat mid face and a short, upturned nose with depressed nasal bridge.

Age of onset   Congenital; progressive myopia starts in the first decade.

Epidemiology   The prevalence of Stickler syndrome in general (types I, II, and III) is about 1:10,000 people. Approximately 50% of Stickler patients have a mutation in *COL2A1* (see below).

Inheritance   Autosomal dominant

Chromosomal location   12q13.11–q13.2

Genes   *COL2A1* (type II collagen α1 chain)

Mutational spectrum   Splice-site mutations and small deletions lead to the introduction of a stop codon.

Effect of mutation   Premature termination of transcription leads to reduced amounts of type II collagen.

Diagnosis   Clinical features may be suggestive for the diagnosis; however, mutation analysis is necessary, since other syndromes can mimic this disease.

Counseling issues   Regular audiologic and ophthalmologic screening in (asymptomatic) affected individuals is advised. Differentiation between types I and II on a clinical basis can be made by eye examination. A membranous vitreous is found in type I and a beaded vitreous in type II. Preventive or early ameliorative treatment (such as surgical retinal reattachment) can be given. Variability of the phenotype is mainly interfamilial; intrafamilial variability is less common.

(b) Stickler Syndrome, Type II

MIM   604841

Clinical features   The clinical features include those described for type I. However, early-onset, moderate to severe SNHI is found in 80%–100% of patients. This SNHI requires the use of a hearing aid. Ocular symptoms are generally less severe than in type I, while orofacial features seem to be more prominent. The congenital myopia is nonprogressive.

Age of onset	Congenital. SNHI is either congenital or with onset in early childhood.
Epidemiology	See type I
Inheritance	Autosomal dominant
Chromosomal location	1p21
Genes	*COL11A1* (type XI collagen α1 chain)
Mutational spectrum	Most mutations affect splicing, leading to in-frame deletions by exon skipping. Amino-acid substitutions are also found.
Effect of mutation	The splice-site mutations result in shortened collagen chains. The amino-acid substitutions lead to structural changes in the collagen chains – for example, shortening of the collagen chain may be found. A disturbance of fibrillar collagen containing COL11A1 can be predicted.
Diagnosis	See type I
Counseling issues	See type I

(c) Stickler Syndrome, Type III

(also known as: autosomal dominant or heterozygous otospondylomegaepiphyseal dysplasia [OSMED]; nonocular Stickler syndrome)

MIM	184840
Clinical features	Similar to those described for type I; however, no ocular symptoms are found since the α2 chain of type XI collagen is not present in the vitreous body of the eye. SNHI is mild to moderate (30–60 dB), shows no progression, and is fully penetrant.
Age of onset	Congenital
Epidemiology	See type I
Inheritance	Autosomal dominant
Chromosomal location	6p21.3

Genes	COL11A2 (type XI collagen α2 chain)
Mutational spectrum	Splice-site mutations lead to exon skipping and a shortened protein.
Effect of mutation	Interference with normal triple helix formation, and thus with the formation of a normal collagen network.
Diagnosis	See type I
Counseling issues	Regular audiologic screening is advised in (asymptomatic) affected individuals.

Weissenbacher–Zweymuller Syndrome

MIM	277610
	The dominantly inherited Weissenbacher–Zweymuller syndrome, Stickler syndrome type III, and homozygous otospondylomegaepiphyseal dysplasia (OSMED) are three syndromes caused by mutations in the COL11A2 (type XI collagen α2 chain) gene. However, COL11A2 may also cause DFNA13 (see p. 32). Another synonym for Weissenbacher–Zweymuller syndrome is "Pierre Robin syndrome with fetal chondrodysplasia". In addition to features of nonocular Stickler syndrome, this syndrome is characterized by neonatal micrognathia and rhizomelic chondrodysplasia with dumbbell-shaped femora and humeri. In later years, regression of bone changes and normal growth are seen. Catch-up growth is observed after 2–3 years. The mutations found cause amino-acid substitutions, which are also predicted to interfere with triple helix formation and collagen fiber structure.

Otospondylomegaepiphyseal Dysplasia

(also known as: OSMED; autosomal recessive or homozygous OSMED)

MIM	215150
	OSMED is a recessive disease caused by homozygosity for a mutation in the COL11A2 (type XI collagen α2 chain) gene. A saddle nose, flat face, hypoplasia of the mandible, and cleft palate are found. SNHI is progressive and severe. Imaging may reveal platyspondyly, carpal bone fusion, limb shortening, and large epiphyses. Degenerative joint disease

is usually more prominent than that found in the various types of Stickler syndrome. Nonsense mutations, deletions, insertions, and splice-site mutations are found, leading to truncated COL11A2 protein, which disturbs collagen fiber formation.

Treacher Collins' Syndrome

(also known as: mandibulofacial dysostosis; Treacher Collins–Franceschetti syndrome)

MIM	154500
Clinical features	Coloboma of the lower eyelid (69%), micrognathia, microtia (77%), ear tags, preauricular sinus, narrowing of the external meatus (36%), conductive hearing impairment (50%), hypoplasia of the zygomatic arches (89%), cleft palate (28%), choanal atresia, mandibular hypoplasia (78%), palatopharyngeal incompetence, macrostomia, and inferior displacement of the lateral canthi of the eyes (89%), usually bilaterally (see **Figure 26**). Eyelashes are sparse medial to the defect of the lower eyelid (see **Figure 27**). Note that coloboma of the upper eyelid is found in Goldenhar syndrome (see p. 87). The middle ear, ossicles, cochlea, and vestibule are often malformed. A conductive hearing impairment is found in >50% of patients, and mixed or sensorineural hearing impairment is found sporadically. Heart defects are occasionally found.
Age of onset	Congenital
Epidemiology	1.0–2.5:100,000
Inheritance	Autosomal dominant
Chromosomal location	5q32–q33.1
Genes	*TCOF1* (treacle)
Mutational spectrum	Nonsense mutations, insertions, deletions, and splice-site mutations have been found. Increasing paternal age seems to predispose to new mutations. No mutational hotspots have been found.
Effect of mutation	The majority of mutations lead to premature termination of translation. Mutations in the *TCOF1* gene lead to a truncated protein, suggesting

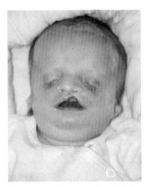

Figure 26. A young girl with Treacher Collins' syndrome. Note the downslanting palpebral fissures, dysplastic auricles, and cleft palate.
Figure courtesy of R Admiraal.

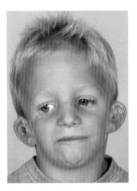 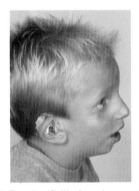

Figure 27. A 4-year-old boy with Treacher Collins' syndrome exhibiting typical features. Ocular characteristics are downslanting palpebral fissures, coloboma of the lower eyelid, and sparse eyelashes medial to the coloboma. Only minor malformation of the auricle is seen in this patient. Hypoplasia of the zygomatic arches and mandible is evident. Figures courtesy of R Admiraal.

that the disease is caused by haploinsufficiency. Treacle may play a role in protein trafficking between the cytoplasm and the nucleolus, which is required during craniofacial development.

Diagnosis   The typical clinical features, radiologic findings, and mutation analysis will lead to the diagnosis. Mutation analysis can be especially helpful in identifying people who are mildly affected or carriers.

Counseling issues	The disease is highly penetrant. There are no indications for genetic heterogeneity of the disorder. There is marked (intrafamilial) phenotypic variation ranging from very mild and difficult to diagnose to perinatally lethal. There is no family history in 60% of cases; these are caused by a *de novo* mutation. Genetic counseling can be difficult due to the presence of very mild symptoms in one of the parents. In cases due to a *de novo* mutation, the risk of a couple having another child with the disease is much lower than for cases where one of the parents carries a *TCOF1* mutation. For the latter cases, the recurrence risk is 50%. Nonpenetrance has even been reported. Radiologic examination of the parents might provide clues for counseling. The detection rate for mutations in the *TCOF1* gene is about 60%. Mutations are often family-specific and there are no mutational hotspots in the gene.

Surgical treatment can involve correcting the coloboma, orbito-malar reconstruction, orthodontic treatment, maxillo-mandibular osteotomies, and refinements such as canthopexy, genioplasty, and auricular correction. Conductive hearing impairment may also be surgically corrected. Besides these interventions, percutaneous titanium implants may be provided to fix prosthetic ears and/or a bone-anchored hearing aid. The aesthetic result of these prosthetic ears can be cosmetically superior to surgical intervention.

Usher Syndrome

The overall prevalence of Usher syndrome is about 3.5–6.2:100,000 people. Prevalence among deaf people has been estimated to be in the range of 3%–6%, and more than 50% of the deaf and blind population is estimated to suffer from Usher syndrome. Three types of Usher syndrome have been differentiated on the basis of clinical criteria, each of which is subdivided (into seven, three, and one subtypes, respectively) on the basis of the genes/loci involved. Additional genetic subtypes are likely to exist, since there are families with subtypes not linked to any of the known loci. Recent studies show that type 2 is the most prevalent form of this syndrome.

(a) Usher Syndrome, Types IA–IG

(also known as: USH1A–1G)

MIM	276900 (USH1A), 276903 (USH1B), 276904 (USH1C), 601067 (USH1D), 602097 (USH1E), 602083 (USH1F), 606943 (USH1G)
Clinical features	Severe to profound congenital hearing impairment, congenitally absent or severely diminished vestibular responses, and pigmentary retinopathy with subsequent loss of vision, starting in the first decade with diminished night vision (see **Figure 28**). Later in childhood, the patient develops decreased peripheral vision that will progress into tunnel vision over decades. Visual acuity also deteriorates with increasing age. Patients usually become blind in late adulthood: 40%, 60%, and 75% at the fifth, sixth, and seventh decade, respectively. Posterior subcapsular cataracts, a surgically treatable condition, has been reported in all types of USH syndrome, although not all patients will develop this. MRI may show a decrease in intracranial volume and in the size of the brain and cerebellum in both USH1 and USH2 patients. Due to their vestibular areflexia, children with all subtypes of USH1 have delayed developmental milestones and will not start walking until the age of about 18–24 months.
Age of onset	Congenital
Epidemiology	USH1A: Mainly reported in families originating from the Poitou-Charentes region of western France. USH1B: The most common of the type 1 syndromes, accounting for 60% of USH1 cases. USH1C: Reported in the French Acadian population and in Lebanese and Muslim families. This accounts for about 5% of all USH1 patients. USH1D: Estimated to constitute about 20% of all USH1 patients. This is the second most common subtype of USH1. USH1E: Reported in a single Moroccan family. USH1F: Estimated to constitute about 10% of all USH1 patients. USH1G: Reported in a Palestinian consanguineous family.
Inheritance	Autosomal recessive

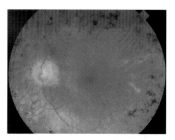

Figure 28. Typical fundoscopic abnormality (left eye) in an Usher patient showing pigmentary retinopathy. Figure courtesy of R Pennings.

Chromosomal location	USH1A: 14q32
	USH1B: 11q13.5
	USH1C: 11p15.1
	USH1D: 10q21–q22
	USH1E: 21q21
	USH1F: 10q21–q22
	USH1G: 17q25.1
Genes	USH1A: Unknown
	USH1B: *MYO7A* (myosin VIIA)
	USH1C: *USH1C* (harmonin)
	USH1D: *CDH23* (cadherin 23)
	USH1E: Unknown
	USH1F: *PCDH15* (protocadherin 15)
	USH1G: *SANS* (scaffold protein containing ankyrin repeats and SAM domain)
Mutational spectrum	USH1A: Unknown
	USH1B: Insertions, deletions, and missense, nonsense, and splice-site mutations are found. Mutations are distributed all over the gene. Mutations in this gene are also reported for DFNB2 (see p. 43) and DFNA11 (see p. 31).
	USH1C: Splice-site mutations, insertions, deletions, and a possible repeat expansion in intron 5 have been found.
	USH1D: Nonsense mutations, insertions, deletions, and splice-site mutations are found. Missense mutations are seen in USH1D in combination with a protein-truncating mutation.

USH1E: Unknown

USH1F: Deletions, nonsense mutations, and splice-site mutations are known, leading to premature termination of protein synthesis.

USH1G: One missense and three frame-shift mutations are known so far.

Effect of mutation   USH1A: Unknown

USH1B: Mutations cause dysfunction of the unconventional myosin encoded by the *MYO7A* gene. Myosin VIIA is an actin-based motor protein that is present in the cochlear hair cell stereocilia, vestibular sensory cells, and retinal pigment epithelium, where it is specifically associated with cilia and microvilli of the apical surface of the cells.

USH1C: *USH1C* encodes a protein called harmonin, which is present in the sensory hair cells of the inner ear and functions in stereocilia organization. The harmonin b isoform is actin-binding. Mutations lead to premature termination of protein synthesis and thus aberrant protein. The effect of the repeat expansion is not yet clear.

USH1D: *CDH23* is expressed in both the cochlea and retina. In mice, *cdh23* mutations have been reported to result in disorganization of inner-ear stereocilia, leading to SNHI. Cadherin 23 has been hypothesized to be part of the (lateral) links between stereocilia, and thereby responsible for the organization of the stereocilia.

USH1E: Unknown

USH1F: As can be deduced from the ames waltzer mouse model, mutations in *PCDH15* lead to abnormal development of stereocilia, resulting in hair cell dysfunction. In the human adult and fetal retina, protocadherin 15 is present in the inner and outer synaptic layers and the nerve fiber layer. The role of the protein in both the cochlea and retina is unknown. In addition to the apical surface of hair cells, the gene is also active in the supporting cells, outer sulcus cells, and the spiral ganglion.

USH1G: The *SANS* gene is probably expressed in both the inner and outer hair cells. The protein associates with harmonin and probably has an important function in the maintenance of the stereocilia bundles.

Diagnosis   Because the phenotypes are similar, the different subtypes are differentiated by linkage and/or mutation analysis. Delayed milestones in children with severe to profound hearing impairment are indicative for

USH1. Linkage analysis can distinguish the genetic subtypes in multicase families, and mutation analysis can distinguish the subtypes for which the corresponding gene has been identified. DNA analyses are almost exclusively performed in research laboratories. Vestibular dysfunction is an expected finding in children with a delayed ability to walk independently. In severely hearing-impaired children, an electroretinogram (ERG) – which records corneal–retinal potentials in response to visual stimuli – may detect retinal disease caused by pigmentary retinopathy in the first decade of life, before the onset of signs and symptoms. Routine ophthalmologic evaluation does not usually show any abnormalities in young children with USH syndrome. In patients with a borderline ERG, this test should be repeated after 18–24 months. Parents of severely hearing-impaired patients with a normal ERG should be informed about the early symptoms of this disease.

Counseling issues   Full penetrance is reported. Early diagnosis is not only important in the scope of family planning, but also in the supporting measures that can be taken. At school, the patient can begin to learn Braille and other supporting skills before vision is severely impaired. Cochlear implants may be very beneficial in this group of patients. There is often high intrafamilial correlation with respect to SNHI. It should be noted that mitochondrial mutations can cause a similar disease, as can viral infections and diabetic neuropathy. Another possible cause is the inheritance of two independent genes, one causing pigmentary retinopathy and the other SNHI.

Carriers of *MYO7A* gene mutations (USH1B) cannot be detected by standard audiometry. Mutations in the *MYO7A* gene can also cause DFNB2 (see p. 43) and DFNA11 (see p. 31). DFNB2 patients may show profound SNHI with onset ranging from birth to 15 years. Patients with DFNA11 may show progressive postlingual SNHI with variable vestibular dysfunction. Remarkably, there seems to be no clear correlation between the type of *MYO7A* mutation and phenotype.

Mutations in the *USH1C* gene are also responsible for DFNB18, a recessive, nonsyndromic SNHI.

In USH1D, different *CDH23* mutations result in a variable retinal phenotype. Mutations in *CDH23* are also seen in patients with atypical USH syndrome (ie, when it is not obvious what type of USH syndrome the patient exhibits). Missense mutations in both alleles of the *CDH23* gene cause recessive SNHI (DFNB12, see p. 48).

DFNB23 may be an allelic variant of USH1F.

(b) Usher Syndrome, Types IIA–IIC

(also known as: USH2A–2C)

MIM	276901 (USH2A), 276905 (USH2B), 605472 (USH2C)
Clinical features	Congenital moderate to severe sloping SNHI – ie, with a downsloping audiogram – normal vestibular responses, and pigmentary retinopathy with onset in the first or second decade. However, intra- and interfamilial variation in this phenotype can be seen. SNHI can be progressive (about 0.5 dB/year for all frequencies), but at a much slower rate than in USH3 patients. Patients can have useful, albeit impaired, vision into middle age. Children with USH2 usually have normal developmental milestones. Posterior subcapsular cataracts are also seen in this type of USH syndrome.
	Audiometric findings in USH2B have shown that USH2B patients also have a downsloping audiogram; however, this is less severe than that reported for USH2A.
Age of onset	Congenital (all types)
Epidemiology	Patients with USH2A are estimated to make up about 80% of the USH2 population. USH2B has been reported in two families, while linkage for USH2C has been reported in nine families.
Inheritance	Autosomal recessive
Chromosomal location	USH2A: 1q41 USH2B: 3p23–p24.2 USH2C: 5q14.3–q21.3
Genes	USH2A: *USH2A* (usherin) USH2B and USH2C: Unknown
Mutational spectrum	Several types of mutations – including missense mutations, nonsense mutations, insertions, deletions, and splice-site mutations – have

been found distributed throughout the protein-coding region of the *USH2A* gene.

Effect of mutation   *USH2A* encodes a protein called usherin, which is a component of basal membranes. The protein is highly expressed in the Bruch's membrane and lens capsule of the retina, and in the stria vascularis of the cochlea. Many mutations lead to premature protein truncation. Aberrant usherin or the absence of usherin are expected to impair basal membrane function in both the retina and cochlea. The 2299delG mutation is a frequently occurring founder mutation. Missense mutations may cause pigmentary retinopathy alone.

Diagnosis   An electroretinogram can be performed, but will be negative in some cases because of the delayed onset of pigmentary retinopathy. Mutation analysis can confirm the diagnosis.

Counseling issues   Mutations are fully penetrant. Patients with USH2 are usually able to communicate orally and attend regular schools. Early diagnosis is not only important in the scope of family planning, but also in the supporting measures that can be taken: eg, accommodations can be made at school. It should be noted that mitochondrial mutations can cause a similar disease, as can viral infections and diabetic neuropathy. Another possible cause is the inheritance of two independent genes, one causing pigmentary retinopathy and the other SNHI.

(c) Usher Syndrome, Type III

(also known as: USH3)

MIM	276902
Clinical features	Progressive hearing impairment, variable vestibular responses, and pigmentary retinopathy with variable onset. Hearing and vision are (near) normal early in life. Hearing progressively deteriorates, although there is much variation in age of onset. Posterior subcapsular cataracts can also be found in these patients.
Age of onset	Extremely variable, but not congenital
Epidemiology	This type of USH syndrome was first seen among Finnish families; therefore, research was mainly conducted in Finland. However, after the

identification of the *USH3* gene (see below), families from Italy, the Netherlands, Spain, Israel, and the US have been reported with mutations in *USH3*. About 2% of all USH patients are thought to suffer from USH3.

Inheritance	Autosomal recessive
Chromosomal location	3q21–q25
Genes	*USH3* (clarin-1)
Mutational spectrum	Nonsense mutations, missense mutations, insertions, and deletions have been identified in the *USH3* gene.
Effect of mutation	The function of this three- or four-transmembrane domain protein (designated clarin-1) is possibly related to sensory synapses.
Diagnosis	Linkage analysis can be performed in multicase families. Mutation analysis can also be performed for the *USH3* gene.
Counseling issues	Full penetrance has been reported. The phenotype may mimic that of the two other types of USH syndrome. Patients with USH3 usually, but not always, communicate orally and attend regular schools. Early diagnosis is not only important in the scope of family planning, but also so that supporting measures can be taken. At school, accommodations can be made. It should be noted that mitochondrial mutations can cause a similar disease, as can viral infections and diabetic neuropathy. Another possible cause is the inheritance of two independent genes, one causing pigmentary retinopathy and the other SNHI.

Waardenburg Syndrome

This is one of the most common types of autosomal dominant syndromic SNHI. The estimated overall prevalence in the general population is 5–10:100,000. The syndrome makes up >2% of children with congenital profound SNHI.

(a) Waardenburg Syndrome, Type I

(also known as: WS1)

MIM	193500

Figure 29. A family suffering from Waardenburg syndrome type I (WS1). The mother and two children show dystopia canthorum. This feature differentiates WS1 from WS2.

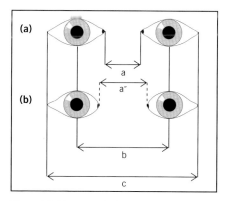

Figure 30. The eyes of (**a**) a nonaffected patient and (**b**) a patient with dystopia canthorum are shown. Note the difference between a and a″. See text for precise definition.

Clinical features Dystopia canthorum (lateral displacement of the inner canthus of each eye) (see **Figures 29** and **30**), pigmentary abnormalities of hair, iris, and skin (leukoderma; about 30%–36% of patients in this type), a white forelock (45%), and heterochromia iridis (15%–31%). The nose seems to show a wide nasal bridge; however, there is no true hypertelorism, but only a widened distance of the canthi medialis, and a normal distance is found between the center of both eyes. A short philtrum, square, prominent jaw, and (radiologically) decreased nasal bone length might be found.

The most serious feature of this syndrome is SNHI, which is found in about 60% of WS1 patients; this is nonprogressive, congenital, and either unilateral or bilateral. Usually, a bilateral loss of >100 dB is found. Pigmentation defects of the skin are correlated with the severity of SNHI. In WS1 patients, the presence of SNHI depends on the genetic background and/or type of mutation. Vestibular hypofunction has frequently been detected in this type. CT scans may reveal abnormal semicircular canals. The fundus of the eye may be completely or partially albinotic. Strabismus has also been reported. The white forelock present at birth may sporadically disappear. A black forelock has also been reported. In some cases, a cleft palate and/or lip have been found; in other patients, spina bifida or lumbosacral myelomeningocele have been observed. Hirschsprung disease and (even more rare) absent vagina and uterine adnexa are reported to be sporadically related to WS1 and WS2.

Age of onset	Congenital
Epidemiology	About 30%–40% of the WS population consists of WS1 patients.
Inheritance	Autosomal dominant
Chromosomal location	2q35
Genes	*PAX3* (paired box protein pax 3)
Mutational spectrum	Several types of mutations are found: nonsense, frame-shift, and splice-site mutations are distributed all over the gene. Missense mutations are concentrated in the DNA-binding regions of the protein.
Effect of mutation	The mutations are thought to destroy the ability of the protein to bind to DNA, and 50% of the normal protein is insufficient for normal differentiation of neural crest cells to melanocytes. This leads to an absence of melanocytes in the skin, hair, eyes, and stria vascularis.

Major criteria	Minor criteria
Congenital SNHI	Hypopigmentation of the skin
Hair hypopigmentation	Medial eyebrow flare
Pigmentary abnormality of the iris	Broad/high nasal root and/or high columella
Dystopia canthorum (W index >1.95)	Premature gray hair (<30 years)
Affected first-degree relative	Hypoplastic alae nasi

Table 1. The major and minor criteria for Waardenburg syndrome, type I.

Diagnosis

The diagnostic criteria for WSI are met if an individual has either two major *or* one major and two minor criteria (see **Table 1**).

The W index formula measures and defines an abnormal distance between the eyes. The W index is calculated as follows:

$$X = (2a - 0.2119c - 3.909)/c$$
$$Y = (2a - 0.2479b - 3.909)/b$$
$$W = X + Y + a/b$$

Where a = inner canthal distance, b = intorpupillary distance, and c = outer canthal distance.

Mutation analysis will usually confirm the diagnosis. Nearly all cases have a mutation in the *PAX3* gene.

Counseling issues

Large intra- and interfamilial variation has been reported for the various characteristics, and there appears to be little correlation between genotype and phenotype. To prevent neural tube defects in WS1 children, pregnant mothers are advised to take folic acid supplements. Reduced penetrance has been reported. A paternal age effect might be present in cases of new mutations (0.4:100,000). In the *PAX* gene, the N47 K mutation leads to craniofacial deafness hand syndrome (MIM 122880), which involves craniofacial abnormalities combined with hand/wrist abnormalities. Translocation between chromosomes 2 and 3, involving the *PAX3* gene, may lead to alveolar rhabdomyosarcoma without any accompanying congenital malformations.

(b) Waardenburg Syndrome, Type II

(also known as: WS2)

MIM	193510
Clinical features	The phenotype is as described for WS1 – except for dystopia canthorum, which is absent in WS2. SNHI is found in about 80% of patients and heterochromia iridis in approximately 50% (see **Figure 31**); this is a higher incidence than in WS1 patients. SNHI, which is often less severe than in WS1, is progressive in about 70% of cases. The incidence of a white forelock (15%–30%) and leukoderma (5%–10%) are lower than in WS1 patients.

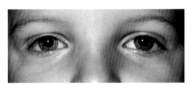

Figure 31. Close-up of the eyes of a Waardenburg type 2 patient. The heterochromia iridis is remarkable.

Age of onset	Congenital
Epidemiology	About 60%–70% of the WS population is made up of WS2 patients.
Inheritance	Autosomal dominant
Chromosomal location	3p14.1–p12.3. Waardenburg syndrome types IIB and IIC have been assigned to chromosomes 1p and 8p23, respectively.
Genes	*MITF* (microphthalmia-associated transcription factor). A small proportion of all WS2 is probably caused by mutations in the *EDNRB* (endothelin receptor B) or *EDN3* (endothelin 3) genes.
Mutational spectrum	Missense mutations, deletions, and splice-site mutations have been found.
Effect of mutation	The mutations are predicted to lead to loss of protein function and thus to the inability of the protein to activate genes involved in development, eg, *MITF*. The mutations result in haploinsufficiency of the MITF protein and disturbance of the normal differentiation of neural crest cells into melanocytes (amongst other cell types) in the stria vascularis.

Diagnosis	See WS1. The diagnosis of WS2 is based on the presence of two major criteria (excluding dystopia canthorum). Minor criteria for WS2 include hypopigmented skin and premature graying of the hair. Mutation analysis can confirm the diagnosis.
Counseling issues	The phenotype of WS2 is variable, even within families. About 10%–20% of WS2 patients have a mutation in the *MITF* gene; others are caused by genes on chromosomes 1p and 8p23 (see chromosomal location). Mutations in this gene, which are predicted to be dominant-negative mutations, have been found in two families with Tietz–Smith syndrome (MIM 103500). One family suffering from WS2 was reported to have a mutation in the *PAX3* gene.

(c) Waardenburg Syndrome, Type III

(also known as: WS3; Klein–Waardenburg syndrome)

MIM	148820
Clinical features	In addition to the features of WS1, patients with WS3 exhibit severe upper limb abnormalities, consisting of hypoplasia of the musculoskeletal system, flexion contractures, carpal bone fusion, and syndactyly. Microcephaly, winged scapulas, mental retardation, and paraplegia may also occur.
Age of onset	Congenital
Epidemiology	Rare
Inheritance	Autosomal dominant or autosomal recessive
Chromosomal location	2q35
Genes	*PAX3* (paired box protein pax 3)
Mutational spectrum	Nonsense and missense mutations have been found, as well as a large deletion including the *COL4A3* gene.
Effect of mutation	The mutations disturb the control of neural crest cell differentiation into melanocytes. PAX3 is a transcription factor, which seems to transactivate the *MITF* promotor.

Diagnosis	The diagnosis can be made on the basis of the clinical characteristics (the major and minor characteristics are the same as for WS1). Mutation analysis can also lead to the diagnosis.
Counseling issues	As with WS1, there seems to be little correlation between genotype and phenotype in these patients.

(d) Waardenburg Syndrome, Type IV

(also known as: WS4; Waardenburg–Hirschsprung disease; Waardenburg–Shah syndrome; Shah–Waardenburg syndrome)

MIM	277580
Clinical features	A combination of the features of recessively inherited WS2 syndrome with Hirschsprung disease. SNHI is seen only sporadically. Hirschsprung disease is defined by the absence of parasympathetic ganglion cells in the distal gastrointestinal tract.
Age of onset	Congenital
Epidemiology	Rare – about 30 families
Inheritance	Autosomal recessive (*EDNRB3* and *EDN3*) or autosomal dominant (*SOX10*)
Chromosomal location	13q22 (*EDNRB*), 20q13.2–q13.3 (*EDN3*), 22q13 (*SOX10*)
Genes	*EDNRB* (endothelin receptor B), *EDN3* (endothelin 3), *SOX10* (SRY-BOX 10)
Mutational spectrum	A missense mutation and deletion have been found in *EDNRB*; a deletion mutation and missense mutations have been observed in *EDN3*; and nonsense mutations have been found to be associated with WS4 in *SOX10*.
Effect of mutation	All mutations cause hypopigmentation, hearing impairment, and lack of enteric ganglia as a result of defective neural crest development.
Diagnosis	Diagnosis is made on the basis of the clinical characteristics. Mutation analysis may also lead to the diagnosis.
Counseling issues	Heterozygous mutations in *EDNRB* cause Hirschsprung disease alone.

Wolfram Syndrome

(also known as: diabetes insipidus, diabetes mellitus, optic atrophy, and deafness [DIDMOAD] syndrome)

MIM	222300
Clinical features	Progressive neurodegeneration, characterized by juvenile insulin-dependent diabetes mellitus and optic atrophy. Hypothalamic diabetes insipidus (73% of patients) and mostly progressive high-frequency SNHI (62%) are additional features responsible for the acronym "DIDMOAD". Other possible features include renal tract abnormalities (58%), such as bladder dystonia, which may cause secondary hydronephrosis; neurologic complications (62%), such as myoclonus and cerebellar ataxia; gastrointestinal dysmotility (24%); and primary gonadal atrophy (7/10 investigated male subjects). Sixty percent of patients also have a variety of psychiatric disorders. These lead to psychiatric submission in 25% of cases. A possible additional suggested feature is gastrointestinal ulceration and a higher bleeding tendency.
Age of onset	Diabetes mellitus and optic atrophy mainly occur in the first decade of life. The second decade is characterized by diabetes insipidus and SNHI. Renal complications mainly occur in the third decade, and neurologic complications in the fourth.
Prevalence	The prevalence is estimated to lie between 1:100,000 and 1:770,000, depending on the type of study.
Inheritance	Autosomal recessive
Chromosomal location	4p16.3
Genes	*WFS1* (wolframin)
Mutational spectrum	Mutations include deletions and insertions, and splice-site, nonsense, and missense mutations. Most of the mutations have been identified in the largest exon (ie, exon 8). No significant hotspots for mutations or clustering of mutations have been identified. About 90% of Wolfram syndrome patients have one or two pathogenic mutations in *WFS1*. A second locus (*WFS2*) for Wolfram syndrome has been mapped to

chromosome 4q22–q24 in three Jordanian families. The causative gene is unknown. Remarkably, no diabetes insipidus occurred in these patients.

Effect of mutation   Mutations in *WFS1* responsible for Wolfram syndrome mainly lead to protein truncation. This is in contrast to the mutations detected for DFNA6/DFNA14 (see p. 23), which is mainly caused by noninactivating mutations in *WFS1*. However, the exact role and function of wolframin remain unknown. It probably plays a crucial role in the survival of specific endocrine and neuronal cells. Biochemical studies show that wolframin is an integral transmembrane protein, predominantly located in the endoplasmic reticulum.

Diagnosis   Juvenile diabetes mellitus and optic atrophy are the minimum diagnostic criteria for this syndrome. Mutation analysis of the *WFS1* gene can be performed to confirm diagnosis in most patients.

Counseling issues   Regular endocrine, urologic, neurologic, ophthalmologic, and audiologic screening is indicated in asymptomatic affected individuals. Wolfram syndrome patients have a median age at death of 30 years and commonly die of central respiratory failure based on brainstem atrophy. Carriers of mutations causing Wolfram syndrome have a significantly higher chance of developing psychiatric disturbances and may have a higher chance of developing diabetes mellitus and hearing impairment. At this moment – outside of symptomatic medical treatment for diabetes mellitus and diabetes insipidus – no treatment modality exists to stop or cure this syndrome.

3. Oxidative Phosphorylation (OXPHOS) Deficiencies

Introduction

In humans, the majority of ATP synthesis occurs via the mitochondrial respiratory chain. Abnormalities in this system can lead to defects in any organ or tissue, but especially in those with the highest energy demand, such as hearing. The protein components of the mitochondrial respiratory chain are encoded in either nuclear or mitochondrial (mt)DNA; thus, genetic defects in either genome can result in impaired oxidative phosphorylation (OXPHOS) (see **Figure 1**). OXPHOS diseases caused by genetic defects in nuclear DNA can exhibit autosomal dominant, recessive, or X-linked inheritance. A genetic defect in mtDNA will exhibit maternal inheritance because mitochondria are only transferred to the next generation via the mother.

Each mitochondrion contains a very large number of copies of double-stranded circular mtDNA. Mutations in mtDNA are not necessarily present in all the copies of the mtDNA. That not all copies of mtDNA are the same – for example, with regard to the presence of a specific mutation – is called "heteroplasmy".

Heteroplasmy and the random distribution of mitochondria to daughter cells can lead to large differences in the ratio of mutated to normal mtDNA between siblings of the same mother and between organs in one individual. This leads to a large variation in clinical presentation between individuals with the same mtDNA mutation. The clinical outcome for a patient with an OXPHOS defect due to a mutation in mtDNA depends on the type of mutation and the percentage of mutant mtDNA in the different organs and tissues. The critical level for the percentage of mutant mtDNA is tissue-specific.

SNHI – either isolated or in combination with other symptoms – is regularly seen in patients with impaired OXPHOS. It has been estimated that approximately 1% of cases of early childhood SNHI are due to OXPHOS defects.

Diagnosis of OXPHOS defects

OXPHOS defects produce a complex group of disorders. They are characterized by a large set of symptoms and symptom overlap between

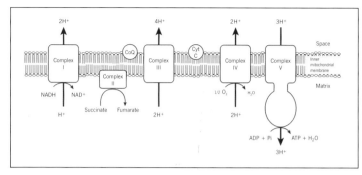

Figure 1. Oxidative phosphorylation. CoQ: coenzyme Q; CytC: cytochrome C.

diseases. Diagnosis of OXPHOS diseases is complicated due to a large phenotypic variation among patients or families with the same disease or mutation. In the following entries, information regarding diagnostic tests is provided, where available. However, diagnosis of this type of disease is complex and should only be performed at centers experienced in the clinical, biochemical, and genetic diagnosis of OXPHOS diseases.

In general, the clinical symptoms can point to an OXPHOS defect. Analysis of body fluids often shows elevated lactate levels. Evaluation of organic acids in urine and amino acids in serum, cerebrospinal liquor, and urine can further increase suspicion of an OXPHOS defect. Following these tests – which will not necessarily be positive – a glucose-tolerance test should be performed. If all of these tests are inconclusive or negative, even though the clinical symptoms strongly suggest impaired mitochondrial function, invasive tests are necessary. These include microscopic and histochemical analysis of a muscle biopsy (musculus vastus lateralis), biochemical analysis of the muscle, and biochemical studies in cultured fibroblasts.

When the tests performed on body fluids are positive, invasive tests are necessary for further investigation. Mutation analysis of nuclear and mtDNA of genes encoding structural and regulatory mitochondrial proteins can confirm the diagnosis and reveal the cause of the disease. Identification of the causative mutation is important for genetic counseling of the family.

MTTS1 gene

(a) Nonsyndromic Maternally Transmitted Hearing Impairment MTTS1-7445

MIM	590080
Clinical features	Mild to severe nonsyndromic hearing impairment, typically affecting the higher frequencies. There is no vestibular dysfunction. In some families, hearing impairment is associated with palmoplantar keratoderma.
Age of onset	Childhood to adolescence
Epidemiology	Four families have been reported.
Inheritance	Maternal
Chromosomal location	Mitochondrial (mt)DNA
Genes	*MTTS1* (mitochondrial transfer [t]RNA serine 1)
Mutational spectrum	7445A>G at the 3′ end of the gene
Effect of mutation	The mutation reduces the processing rate of the tRNA precursor and consequently the rate of protein synthesis. The level of NADH dehydrogenase subunit 6 (ND6) messenger (m)RNA is also reduced. ND6 is a component of mitochondrial complex I, and cell lines with the mutation exhibit a phenotype suggestive of defective OXPHOS.
Diagnosis	This disease is suggested by a maternal pattern of inheritance of nonsyndromic SNHI; however, penetrance is variable. The final diagnosis can be reached via muscle biopsy or mutation analysis.
Counseling issues	There is no correlation between the degree of hearing impairment and the degree of heteroplasmy, and as yet unidentified factors may play a crucial role. Counseling is complicated due to the complexity of mtDNA inheritance, tissue variation in heteroplasmy, and problems with correlating the genotype and phenotype. Penetrance of the disorder is incomplete and varies from 50%–100% in the reported families.

(b) Oligosyndromic Maternally Transmitted Hearing Impairment MTTS1-7472

MIM	590080
Clinical features	Hearing impairment presents in the higher frequencies with tinnitus and vertigo. Patients may show vestibular hyperreactivity, with special susceptibility to motion sickness. Neurologic symptoms – such as ataxia, dysarthria, muscle fatigue, focal myoclonus, co-ordination problems, and late-onset diabetes – occur at a later stage of the disease. This mutation is also found in some subtypes of cytochrome C oxidase deficiencies that are associated with ataxia, myoclonus epilepsia, and mental retardation.
Age of onset	Variable, but not later than about 45 years of age for hearing loss. Neurologic symptoms occur earlier.
Epidemiology	Six families are known.
Inheritance	Maternal
Chromosomal location	Mitochondrial DNA, nucleotides 7445–7516
Genes	*MTTS1* (mitochondrial transfer [t]RNA serine 1)
Mutational spectrum	Insertion of one nucleotide, m.7472insC
Effect of mutation	The mutation leads to complex I and IV OXPHOS deficiencies. The extra cytosine alters the tertiary structure of the *MTTS1* gene at the 3′ end, leading to a decreased level of tRNA(ser)(ucn) and mild respiratory impairment (shown in cultured cells).
Diagnosis	This mutation should be strongly suspected in families with a clear maternal pattern of inheritance of hearing impairment with "episodes of vertigo" and tinnitus, with or without the presence of minor neurologic complaints such as focal myoclonus and ataxia. The final diagnosis can be reached via muscle biopsy or mutation analysis. Large mitochondria and paracrystalline inclusions can be present in the mitochondria (see **Figure 2**).

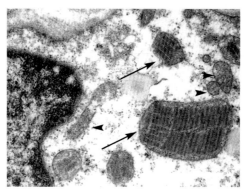

Figure 2. The arrowheads indicate normal mitochondria. Enlarged mitochondria with paracrystalline inclusions and "parking lot" arrangements are shown (arrows).

Counseling issues	The degree of hearing impairment does not correlate with the degree of heteroplasmy, and unidentified factors might play a crucial role. To date, sporadic cases with this mutation have not been described. Three members of one of the families with the m.7472insC mutation developed hearing impairment after the use of aminoglycosides. Penetrance of the disease is almost complete.

(c) Nonsyndromic Maternally Transmitted Hearing Impairment MTTS1-7510

MIM	590080
Clinical features	Moderate to profound hearing impairment with a gently downsloping audiogram and maternal inheritance.
Age of onset	Highly variable, from childhood to adult.
Epidemiology	One family has been reported.
Inheritance	Maternal
Chromosomal location	Mitochondrial (mt)DNA

Genes	MTTS1 (mitochondrial transfer [t]RNA serine 1)
Mutational spectrum	The nucleotide substitution m.7510T>C
Effect of mutation	The mutation is predicted to disrupt base-pairing in the acceptor stem of tRNA, and thus it is likely that the amount of tRNA is affected, which will impair mitochondrial protein synthesis (cf, the m.7445T>C and m.7472insC mutations, p. 130 and p. 131, respectively).
Diagnosis	The diagnosis should be suspected in families with a maternal pattern of inheritance of nonsyndromic hearing impairment. A final diagnosis may be reached via muscle biopsy or mutation analysis.
Counseling issues	There appears to be no correlation between hearing impairment and heteroplasmy.

(d) Syndromic Maternally Transmitted Hearing Impairment MTTS1-7512 in a MELAS/MERRF Overlap Syndrome

MIM	590080
Clinical features	Variable and fluctuating hearing impairment with mental disturbances in a MELAS (mitochondrial encephalomyopathy, lactic acidosis, stroke-like episodes)/MERRF (myoclonic epilepsy associated with ragged-red fibers) overlap syndrome, with truncal ataxia and muscle atrophy as the main clinical features. This mutation can also cause a combination of hearing impairment, ataxia, and dementia.
Age of onset	25–43 years
Epidemiology	Two families have been identified.
Inheritance	Maternal inheritance and isolated cases
Chromosomal location	Mitochondrial (mt)DNA
Genes	MTTS1 (mitochondrial transfer [t]RNA serine 1)
Mutational spectrum	m.7512T>C

Effect of mutation	The mutation is predicted to disrupt base-pairing in the acceptor stem of tRNA, and thus it is likely that the amount of tRNA is affected (*cf*, the m.7445T>C and m.7472insC mutations, p. 130 and p. 131, respectively). Mitochondrial protein synthesis is impaired.
Diagnosis	This syndrome should be suspected in families with a maternal pattern of inheritance of nonsyndromic hearing impairment and associated neurologic symptoms that are comparable with those of MELAS (p. 137), MERRF (p. 141), or an overlap syndrome.
Counseling issues	Young patients, especially those with syndromic disease, are usually seen by pediatricians. Neurologists, ophthalmologists, and (sometimes) specialists in internal diseases are involved in the treatment of these patients. Treatment, which consists mainly of supportive measures, should be provided in specialized centers.

Nonsyndromic Maternally Transmitted Hearing Impairment caused by Mutations in the *MTRNR1* Gene

MIM	561000
Clinical features	Nonsyndromic, maternally transmitted symmetric hearing impairment, affecting the high frequencies most severely. In many cases, hearing impairment progresses to profound with increasing age. No vestibular dysfunction is seen. Mutations in the *MTRNR1* gene can be associated with acute aminoglycoside-induced deafness (AAID) following the administration of low doses of aminoglycosides.
Age of onset	Variable, but not later than approximately 45 years of age.
Epidemiology	This type of hearing loss has an unknown incidence. The highest prevalence is in south-east Asia and the Arab–African region. In Spain, the m.1555A>G mutation (see below) was found in 19 out of 70 families with hearing impairment. The frequency of the two additional mutations (m.1095T>C and m.961delTinsC) is unknown.
Inheritance	Inheritance is maternal, with some isolated cases also being reported. In one family, the pattern of inheritance fitted a two-locus model, combining autosomal recessive and maternal inheritance.

Chromosomal location	Mitochondrial (mt)DNA
Genes	*MTRNR1* (mitochondrial 12S ribosomal [r]RNA)
Mutational spectrum	A nucleotide substitution at mitochondrial nucleotide 1555 (m.1555A>G). Also, m.1095T>C and m.961delTinsC.
Effect of mutation	The mutation at nucleotide 1555 alters an aminoglycoside binding site on rRNA, causing aminoglycoside-induced reduction of protein synthesis. The 1095 mutation destroys the stem-loop structure of the RNA, leading to reduced mitochondrial protein synthesis and reduced cytochrome C oxidase activity. The specific effect of the m.961delTinsC mutation at the molecular level is unknown.
Diagnosis	Most individuals with AAID experience hearing impairment within 3 months of "nontoxic" aminoglycoside administration. There is no correlation between the degree of heteroplasmy and hearing impairment.
Counseling issues	The mutations should be suspected in families with a maternal pattern of inheritance with variable presentation of nonsyndromic hearing impairment. In individuals with AAID, testing should also be performed for mutations in the *MTRNR1* gene. The m.1555A>G mutation is responsible for only a minority of patients with aminoglycoside-induced SNHI. A modifier gene of the phenotype of the m.1555A>G mutations is present on chromosome 8.

Kearns–Sayre Syndrome

(also known as: KSS; chronic progressive external ophthalmoplegia [CPEO] with ragged-red fibers)

MIM	530000
Clinical features	Ophthalmoplegia, atypical retinitis pigmentosa, mitochondrial myopathy, and one of the following: cardiac conduction defect, cerebellar syndrome, or an elevated level of cerebrospinal fluid protein. Characteristics of CPEO can vary from isolated extraocular eye muscle involvement to "CPEO-plus", in which a variety of the clinical manifestations seen in KSS are also present (nb, KSS, CPEO, and CPEO-plus are variations of

the same syndrome). In addition, patients with KSS and CPEO-plus have a variable combination of hearing impairment, dementia, seizures, hypertrophic and dilated cardiomyopathy, and cardiac arrhythmias. Other characteristics include diabetes mellitus, hypoparathyroidism, renal failure, respiratory failure, gastrointestinal mobility disturbances, mitochondrial myopathy, lactic acidemia, and sensory and motor neuropathies. SNHI occurs in at least 50% of patients, and presents mainly and initially in the high frequencies.

Age of onset Onset is before the age of 20 years in KSS, and after 20 years in CPEO and CPEO-plus.

Epidemiology Unknown, although many patients have been described.

Inheritance Most cases are sporadic, but clear maternal inheritance has been described in a few families. Cases of autosomal recessive inheritance have also been reported; autosomal dominant inheritance has been reported for CPEO.

Chromosomal location Mitochondrial (mt)DNA, chromosomes 3p14.1–p21.2 and 10q23.3–q24.3, and unknown loci

Genes Several mitochondrial genes can be involved (see below), and as yet unknown nuclear genes.

Mutational spectrum Mitochondrial DNA rearrangements (deletions and duplications in combination with a deletion) have been found in about 80% of KSS cases, 70% of CPEO-plus cases, and 40% of CPEO cases. In particular, a 4,977 base-pair deletion encompassing the genes for several mitochondrial enzyme complexes is frequently found. Other deletions vary in size (2–7 kb). In one case, the mutation responsible for MELAS (mitochondrial encephalomyopathy, lactic acidosis, stroke-like episodes) syndrome was identified. In patients with CPEO, point mutations in mitochondrial transfer (t)RNA genes are found (*MTTL2* [tRNA, mitochondrial, leucine, 2], *MTTI*, *MTTN* [tRNA, mitochondrial, asparagine]).

Effect of mutation The mutations affect genes that are essential for protein synthesis in mitochondria, and therefore impair protein synthesis and thus mitochondrial function. In patients with a genetic defect in mtDNA, the stochastic distribution of mutant molecules during development determines the symptoms.

Diagnosis	Any case presenting with external ophthalmoplegia and hearing impairment before the age of 20 years should raise suspicion. In KSS, the amount of mutant mtDNA compared with wild-type DNA is high, mainly in muscle and the central nervous system.
Counseling issues	Symptoms can be highly variable within the same family, ranging from asymptomatic to severe. Because most cases are sporadic, there are no special genetic counseling issues. The risk of transmission of a deletion is low. The 4,977 base-pair deletion is also frequently found in patients with Pearson syndrome (an OXPHOS syndrome that mainly affects the bone marrow; MIM 557000); children who survive Pearson's syndrome frequently develop KSS. Some patients demonstrate an overlap with other OXPHOS diseases, such as MELAS. Mitochondrial DNA rearrangement syndromes seem to represent a phenotypic continuum, with isolated diabetes mellitus and SNHI being the mildest presentation, CPEO and KSS intermediate, and Pearson's syndrome the most severe phenotype.

MELAS Syndrome

(also known as: mitochondrial encephalomyopathy, lactic acidosis, stroke-like episodes)

MIM	540000
Clinical features	Progressive neurodegeneration, stroke-like episodes, and mitochondrial myopathy with ragged-red muscle fibers. Other manifestations include headache, vomiting, lactic acidosis, myalgias, ophthalmoplegia, diabetes mellitus type 2, pigmentary retinopathy, dementia, ataxia, renal disease, and cardiac conduction defects. SNHI is found in approximately 30% of cases.
Age of onset	5–15 years, but earlier and later onset can be seen
Epidemiology	Prevalence is 16:100,000 in the Finnish population.
Inheritance	Maternal inheritance and isolated cases
Chromosomal location	Mitochondrial (mt)DNA

Genes	*MTTL1* (transfer [t]RNA, mitochondrial, leucine, 1), *MTND1* (complex 1, NADH dehydrogenase, subunit 6), *MTND4*, *MTCO3* (cytochrome C oxidase III), *MTTV* (tRNA, mitochondrial, valine)
Mutational spectrum	The nucleotide substitutions m.3243A>G (80%) or m.3271T>C (7.5%) are found in the *MTTL1* gene. Mutations in the other genes are missense mutations.
Effect of mutation	Mutations that affect the *MTND1*, *MTND4*, and *MTCO3* genes directly impair OXPHOS. The mutations in the *MTTL1* and *MTTV* genes impair protein synthesis and thereby lead to impaired mitochondrial functioning. The m.3243A>G mutation also changes a nucleotide at the binding site for a transcription termination factor, and might therefore interfere, for example, with the processing of 16SrRNA or tRNA(leu)(uur). The overall result is impaired protein synthesis in the mitochondria.
Diagnosis	Recurrent strokes are the clinical hallmark of this disease and should lead to testing for MELAS syndrome. Vascular pathology is important in the pathogenesis of stroke and angiopathy is sometimes visible in the skin as purpura.
Counseling issues	Numerous different phenotypes of the m.3243A>G mutation have been reported. Several syndromes overlap with MELAS. Counseling is complicated by the complexity of mtDNA inheritance, tissue variation in heteroplasmy, and difficulties in correlating genotype and phenotype. In individuals with late-onset diabetes mellitus and early-onset high-frequency hearing impairment (MIM 520000), the MELAS mutation is also present. In one sporadic case with a severe phenotype, the m.14453G>A mutation was found in the mitochondrial *ND6* gene. This, in combination with two other mitochondrial mutations (m.5628T>C in *MTTA* and m.13535A>G in *MTND5*), might have caused the severe phenotype.

Leigh Syndrome

(also known as: subacute necrotizing encephalopathy)

MIM	256000
Clinical features	The clinical manifestations are variable, but cranial nerve abnormalities (optic atrophy and ophthalmoplegia), hypotonia, and ataxia are the

primary clinical signs. Other symptoms include weakness, respiratory distress (dyspnea), apnea and Cheyne–Stokes breathing, SNHI, polyneuropathy, and developmental delay or regression. Hypertrophic cardiomyopathy can occur, and there may be liver dysfunction and retinitis pigmentosa.

Age of onset	Usually infancy or early childhood
Epidemiology	1:40,000
Inheritance	Maternal, X-linked, and autosomal recessive inheritance can be seen. Isolated cases have also been described.
Chromosomal location	Mitochondrial (mt)DNA and loci on 19p13, 11q13, 9q32.2–q34.11, 9q34, 5q11.1, 5p15, 2q33, and Xp22
Genes	*MTATP6* (ATP synthase 6), *MTTV* (transfer [t]RNA, mitochondrial, valine), *SURF1* (surfeit-1), *PDHA1* (pyruvate dehydrogenase complex, E1-α polypeptide 1), *BCS1L* (BCS1-like), *NDUFV1* (NADH-ubiquinone oxidoreductase flavoprotein 1), *NDUFS4*, *NDUFS7*, *NDUFS8*, and *SDHA* (succinate dehydrogenase complex, subunit A, flavoprotein). These genes encode subunits of several complexes of the respiratory chain. In addition, the *MTTL1* (tRNA, mitochondrial, leucine, 1) and *MTTK* (tRNA, mitochondrial, lysine) genes may be involved.
Mutational spectrum	The mitochondrial mutations lead to amino-acid substitutions. Mitochondrial deletions have also been described. The mutations in nuclear genes (all but *MTATP6*, *MTTV*, *MTTL1*, and *MTTK*) are missense or splice-site mutations, insertions, and deletions. The mitochondrial mutations m.8993T>G (*MTATP6*), m.8993T>C (*MTATP6*), and m.8344 A>G (*MTTK*) are the most common mutations in children with Leigh syndrome.
Effect of mutation	Mutations that cause Leigh syndrome lead to a severe defect in ATP production. Defects in complexes I, IV, and V of OXPHOS are important causes of Leigh syndrome. In addition, defects in complexes II and III are seen.
Diagnosis	The presence of the clinical features can suggest the diagnosis. T2-weighted MRI reveals hyperintense signals in the basal ganglia, cerebellum, or brainstem. Muscle biopsy can show altered fiber type

and size, (rarely) ragged-red fibers, and (occasionally) abnormal mitochondrial ultrastructure.

Counseling issues	The mean duration of illness until death is about 5 years. For patients with mutations in mtDNA, there is a strong correlation between mutational load (heteroplasmy) and symptoms. Leigh syndrome can also be caused by defects in a variety of metabolic pathways, such as the pyruvate dehydrogenase complex; however, OXPHOS defects are the most common cause of Leigh syndrome. Mitochondrial mutations at position 8993 are also seen in patients with NARP (neuropathy, ataxia, and retinitis pigmentosa; MIM 551500), MERRF (myoclonic epilepsy associated with ragged-red fibers, p. 141), and MELAS (mitochondrial encephalomyopathy, lactic acidosis, stroke-like episodes, previous entry).

Leber's Hereditary Optic Atrophy

(also known as: Leber's hereditary optic neuropathy [LHON])

MIM	535000
Clinical features	Acute or subacute progressive bilateral central vision loss, leading to a central scotoma with peripapillar telangiectasia and vascular tortuosity. Most families only manifest optic atrophy. Apart from these indications, the most common clinical sign is dystonia, frequently with pseudo-bulbar syndrome, short stature, and myopathic signs. In addition, cardiac conduction defects, nonspecific hearing impairment, tremors, sensory neuropathy, ataxia, and extrapyramidal signs have occasionally been described. Multiple sclerosis-like optic neuritis can be present in LHON family members.
Age of onset	25–35 years, mean 27 years
Epidemiology	Unknown. Many isolated cases have been reported.
Inheritance	Maternal inheritance or isolated cases
Chromosomal location	Mitochondrial (mt)DNA

Genes	Several mitochondrial genes, including: *MTND1* (complex 1, NADH dehydrogenase, subunit 6), *MTND2*, *MTND4*, *MTND5*, *MTND6*, and *MTATP6* (ATP synthase 6).
Mutational spectrum	Nucleotide substitutions leading to missense mutations have been detected. The most frequent mutations are m.11778G>A (*MTND4*), m.3460G>A (*MTND1*), and m.14484 T>C (*MTND6*). These mutations are present in >90% of European and Asian LHON cases.
Effect of mutation	All mutations cause a defect in OXPHOS complexes I, III, and IV, acting either independently or in association with each other. The overall result of this is mitochondrial dysfunction.
Diagnosis	Acute loss of central vision should suggest the diagnosis of LHON. There are no reliable metabolic, biochemical, or neuroradiologic tests for LHON. Mutation analysis is efficient because of the high percentage of cases (>90%) with the three most common mutations.
Counseling issues	Counseling is difficult due to the complexity of mtDNA inheritance, tissue variation in heteroplasmy, and difficulties in correlating genotype and phenotype. In most patients, the pathogenic mutation is homoplasmic in blood (ie, all copies of the mitochondrial DNA carry this mutation).

MERRF Syndrome

(also known as: myoclonic epilepsy associated with ragged-red fibers)

MIM	545000
Clinical features	Progressive myoclonic epilepsy, mitochondrial myopathy with ragged-red fibers, and progressive dementia. SNHI and ataxia are commonly seen. An accumulation of lipomas in the anterior and posterior cervical region is often seen in MERRF patients.
Age of onset	Late childhood to adulthood
Epidemiology	Unknown. Many families have been reported.
Inheritance	Maternal
Chromosomal location	Mitochondrial (mt)DNA

Genes	*MTTK* (transfer [t]RNA, mitochondrial, lysine), *MTTL1* (tRNA, mitochondrial, leucine, 1)
Mutational spectrum	The nucleotide substitution m.8344A>G (*MTTK*) is responsible for 80%–90% of MERRF cases. The substitution m.8356T>C is also found in the same gene. The m.3256C>T substitution is found in the *MTTL1* gene.
Effect of mutation	Often, combined cytochrome C oxidase (complex IV) and complex I deficiencies occur due to a translational defect of mtDNA-encoded genes. Other defects include: complex II; cytochrome B (complex III) or cytochrome C; combined complex III and IV defects; or coenzyme Q.
Diagnosis	The clinical picture should suggest the diagnosis of MERRF syndrome. Positron emission tomography and ^{31}P-nuclear magnetic resonance (NMR) imaging can detect impaired energy metabolism in the brain; ^{31}P-NMR can also detect this in muscle. MRI of the brain can reveal abnormalities such as cerebral or cerebellar atrophy, cerebellar hyperintense lesions, basal ganglia hyperintense lesions, hyperintense white matter, or pyramidal tract lesions.
Counseling issues	Counseling is complicated due to the complexity of mtDNA inheritance, tissue variation in heteroplasmy, and difficulties in correlating genotype and phenotype. MERRF can be distinguished from MELAS (mitochondrial encephalomyopathy, lactic acidosis, stroke-like episodes, p. 137) by the absence of strokes. Features of both MELAS and MERRF can be seen in patients with MERRF/MELAS overlap. MERRF/CPEO (chronic progressive external ophthalmoplegia) overlap also exists.

MNGIE Syndrome

(also known as: myoneurogastrointestinal encephalopathy syndrome)

MIM	603041
Clinical features	Diarrhea, malabsorption, and weight loss. Other typical features are progressive external ophthalmoplegia; mitochondrial myopathy with ragged-red fibers; peripheral, motor, and sensory neuropathy; and dementia with progressive leukodystrophy. SNHI presents at around 30 years of age.

Age of onset	Between the first and fifth decades
Epidemiology	Unknown
Inheritance	Autosomal recessive; sporadic cases have also been reported
Chromosomal location	22q13.32–qter
Genes	*ECGF1* (endothelial cell growth factor 1, thymidine phosphorylase)
Mutational spectrum	Nucleotide substitutions (missense mutations), small insertions and deletions, and splice-site mutations are found. These lead to amino-acid deletions or protein truncation.
Effect of mutation	The mutations lead to aberrant thymidine metabolism (increased plasma thymidine). This might cause impaired replication and/or maintenance of mitochondrial (mt)DNA, leading to the multiple deletions of mtDNA seen in these patients.
Diagnosis	Growth deficiency – as a result of diarrhea and malabsorption – is the most striking characteristic that can lead to the diagnosis. The diagnosis can be suspected by urine/blood thymidine measurements and confirmed by mutation analysis.
Counseling issues	In many cases, the outcome is severe disability and death.

Diabetes–Deafness Syndrome, Maternally Transmitted

MIM	520000
Clinical features	In most cases, SNHI precedes diabetes mellitus. SNHI presents in the high frequencies. There is rapid progression towards severe bilateral hearing loss.
Age of onset	The onset of SNHI varies between 20 and 40 years of age. The onset of diabetes mellitus is in the second to fourth decades.
Epidemiology	About 5% of Japanese diabetics carry this mutation.
Inheritance	Both isolated cases and maternal inheritance have been reported.

Chromosomal location	Mitochondrial (mt)DNA
Genes	Rearrangements affect a large number of mitochondrial genes. *MTTL1* (transfer [t]RNA, mitochondrial, leucine, 1) and *MTTK* (tRNA, mitochondrial, lysine) are also affected.
Mutational spectrum	Complex rearrangements of mtDNA have been described, including deletions, duplications, and a triplication. Nucleotide substitutions m.3243A>G (*MTTL1*) and m.8296A>G (*MTTK*) have been found in the tRNA genes.
Effect of mutation	Both mtDNA rearrangements and the point mutations affect genes that are essential for mitochondrial protein synthesis. These mutations therefore impair protein synthesis and thus mitochondrial function.
Diagnosis	In most cases, diabetes presents later than hearing impairment. A sensitive tool is to demonstrate defects in glucose-related insulin secretion. Genetic testing can be performed.
Counseling issues	The risk of transmission of a deletion is low, but not negligible. Mitochondrial DNA rearrangement syndromes seem to represent a phenotypic continuum, with isolated diabetes mellitus and SNHI being the mildest presentation, chronic progressive external ophthalmoplegia and Kearns–Sayre syndrome (see p. 135) intermediate, and Pearson syndrome (MIM 557000) the most severe phenotype. There is also an association between SNHI and diabetes in Wolfram syndrome (see p. 125) and multiple synostoses syndrome 1 (MIM 186500) – however, these syndromes are known to be caused by a mutation in the nuclear genome.

4. Inherited Diseases in Rhinology

Kallmann Syndrome, Types I–III

(also known as: KAL1, KAL2, KAL3)

Introduction	Kallmann syndrome exists in three types: KAL1, KAL2, and KAL3. All three have an estimated prevalence of 1:10,000 males and 1:50,000–70,000 females. Most cases are sporadic; familial cases are rare. The *KAL1* gene is responsible for only a minority of cases.
	The three types are mainly differentiated by their mode of inheritance: KAL1, KAL2, and KAL3 show X-linked, autosomal dominant, and autosomal recessive inheritance, respectively. Phenotypic differences mainly consist of the presence of a more variable phenotype in KAL2 and the presence of midline defects in KAL3.
	SNHI – mostly mild, congenital, and bilateral – is found in approximately 20% of patients and can be asymmetric. Mixed hearing impairment has also been reported. Vestibular responses may be absent. A CT scan may reveal abnormal semicircular canals (the vestibule can be seen, but there is no identifiable normal semicircular canal structure) and an abnormal internal acoustic canal (this is unusually slender, and the lateral termination may be unusually high in relation to the head of the malleus and incus).
MIM	308700 (KAL1), 147950 (KAL2), 244200 (KAL3)
Clinical features	KAL1: Males usually show anosmia and have small genitalia with sparse sexual hair as a result of a deficiency in the release of hypothalamic gonadotropin-releasing hormone (GnRH). Cryptorchidism and infantile testis may also be found. Females show partial or complete anosmia, and sporadically show hypogonadism. Anosmia is a result of olfactory lobe agenesis. Other, rare features are eye movement abnormalities, gaze-evoked horizontal nystagmus, spatial-attention abnormalities, and cleft lip and palate. Some patients show unilateral renal agenesis (35%) (bilateral renal agenesis is usually fatal), bimanual synkinesia (75%), pes cavus, a high-arched palate, gynecomastia (in males), or cerebellar ataxia. Bilateral SNHI has been reported in several patients (about 20% in all three types).
	KAL2: Variable. Hypogonadism and anosmia are usually found. Other features include mental retardation, choanal atresia, congenital heart defects (such as coarctation of the aorta, double-outlet right ventricle,

and malpositioning of great arteries), cryptorchidism, perceptive hearing impairment, and short stature (height ≥ 2 standard deviations below the mean for age).

KAL3: Hypogonadotropic hypogonadism, anosmia, and midline cranial anomalies. Midline cranial anomalies may consist of a cleft lip, cleft palate, and imperfect fusion of the facial midline, resulting in a median cleft or, for example, holoprosencephaly. Hypotelorism and unilateral renal agenesis are also sometimes seen.

Age of onset	Congenital. KAL1 is diagnosed at a mean age of approximately 25 years.
Epidemiology	1:10,000 males and 1:50,000–70,000 females
Inheritance	X-linked (KAL1); autosomal dominant (KAL2); autosomal recessive (KAL3). Autosomal dominant and X-linked inheritance have also been described for KAL3 in some cases.
Chromosomal location	Xp22.3 (KAL1); 8p11.2–p11.1 (KAL2); genetic heterogeneity is likely for KAL2. The chromosome location for KAL3 is unknown.
Genes	*KAL1* (anosmin-1; KAL1); *FGFR1* (fibroblast growth factor receptor 1; KAL2); unknown for KAL3
Mutational spectrum	Missense and nonsense mutations, frame-shift mutations due to small insertions or deletions, splice-site mutations, and deletions of part or all of the gene (*KAL1*). Missense, nonsense, and splice-site mutations have been described for *FGFR1*, as well as deletions and a small insertions. Unknown for KAL3.
Effect of mutation	KAL1: The majority of mutations lead to absence or truncation of anosmin-1, an extracellular matrix component. Mutations lead to defects in migration (from the olfactory epithelium to the hypothalamus) and target recognition of olfactory neurons and GnRH-secreting cells, causing anosmia/hyposmia and endocrine disorders.
	KAL2: The described mutations in the *FGRF1* gene can be expected to be loss of function mutations. Apparently, the development of the olfactory bulb is sensitive to a reduced dosage of FGFR1. It has been hypothesized that FGFR1 is involved in anosmin-1 signaling.
	KAL3: Unknown

Diagnosis	Patients with anosmia or hyposmia should be examined carefully for other symptoms of the disorder. Laboratory tests might reveal deficiencies in GnRH, follicle-stimulating hormone, or luteinizing hormone. A definitive diagnosis can be reached by mutation analysis. A CT or MRI scan may identify cerebral and inner-ear abnormalities – these can be important diagnostic features, since it is often difficult to test olfaction and the hypothalamo-pituitary–gonadal axis in young patients.
	KAL1 and KAL2 are differentiated by their mode of inheritance and the presence of clinical features such as short stature, congenital heart defects, and mental retardation in KAL2. Midline cranial anomalies differentiate KAL3 from KAL1 and KAL2. Consanguinity will support recessive inheritance.
Counseling issues	An early and accurate diagnosis is important so that GnRH replacement therapy can be given. Families with this disorder exhibit a stable phenotype. Abdominal ultrasound may reveal renal aplasia in KAL1 and KAL3. Patients with a contiguous gene syndrome that includes the *KAL1* gene exhibit ichthyosis, mental retardation, chondrodysplasia punctata, short stature, and ocular albinism. The autosomal recessive form of KAL3 shows variable expression.
	Hearing impairment is mild, so not all patients need a hearing aid at a younger age. However, many will need a hearing aid fitting when they grow older. GnRH replacement therapy usually corrects short stature. Patients who are not given GnRH replacement therapy are often infertile; even if therapy is given, fertility might be absent.

Congenital Anosmia

MIM	107200
Clinical features	Patients present with either anosmia or hyposmia. These are the only clinical features. The mechanism of the anosmia is unknown, but bilateral aplasia of the olfactory bulbs has been demonstrated in some patients with isolated congenital anosmia.
Age of onset	Congenital
Epidemiology	About 5–10 families
Inheritance	Autosomal dominant

Chromosomal location	Unknown
Genes	Unknown
Mutational spectrum	Unknown
Effect of mutation	Unknown
Diagnosis	Congenital anosmia can be diagnosed in cases of anosmia, in the absence of other symptoms, where dominant inheritance is found.
Counseling issues	See Kallmann syndrome types I–III (previous entry) for possible accompanying anomalies. If these are found, then the diagnosis must be revised. Correction of this disorder is not possible.

Rendu–Osler–Weber Syndrome, Types I–III

(also known as: Osler–Rendu–Parkes–Weber syndrome; hereditary hemorrhagic telangiectasia)

MIM	187300 (type I), 600376 (type II), 601101 (type III)
Clinical features	Multiple telangiectases of the skin and mucosa resulting in recurrent bleeding, and arteriovenous fistulae at various sites. Telangiectases may also occur on the tongue and lips (60% of patients; see **Figure 1**); face and ears (35%); conjunctiva (35%); and fingers, toes, and nail beds (40%). In most cases, these are of minor importance, and mainly give rise to cosmetic complaints. Vascular dysplasia can lead to recurrent (sometimes severe) epistaxis (95%) and gastrointestinal bleeding. Gastrointestinal hemorrhage is found in about 16% of patients, half of whom require blood transfusion. Visceral involvement with arteriovenous malformations may affect the lungs (5%–15%), liver (8%), kidneys, bladder, and brain (at least 5%) – their presence can be screened for. Pulmonary arteriovenous malformations are found in up to 20% of patients with type I, although there is a much lower incidence in types II and III (type III is intermediate between types I and II). These may cause heart failure leading to polycythemia and clubbing. Paradoxical emboli can cause infarction or abscess formation in the brain and elsewhere. Vascular malformations in the liver are a possible feature of type III.

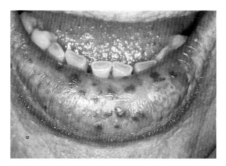

Figure 1. Multiple telangiectases of the lip and tongue in a patient with Rendu–Osler–Weber syndrome. Figure courtesy of I van der Waal.

Age of onset	The onset of epistaxis (often the first symptom) is, on average, at around 12 years of age. Ninety percent of patients are affected by the age of 21 years.
Epidemiology	Estimates vary from 1:8,345 to 1:100,000. The variable expression makes underestimation likely.
Inheritance	Autosomal dominant
Chromosomal location	9q34.1 (type I); 12q11–q14 (type II); unknown for type III
Genes	*ENG* (endoglin; type I); *ALK1* (activin receptor-like kinase-1; type II); unknown for type III
Mutational spectrum	Type I mutations include missense, nonsense, and splice-site mutations, small insertions and deletions, and complex rearrangements. Missense, nonsense, and frame-shift mutations have been found in type II. Unknown for type III.
Effect of mutation	Type I: Vascular dysplasia caused by haploinsufficiency of the *ENG* gene leads to telangiectases (only the endothelial wall is present) and arteriovenous malformations of the skin, mucosa, and viscera. The mutated allele of the gene leads to mRNA that is either rapidly degraded or encodes a shorter or aberrant protein. Therefore, there is insufficient functional protein in some parts of the body or at some parts of development (haploinsufficiency). Endoglin is a transforming growth factor (TGF)-β binding protein. TGF-β regulates various processes of

endothelial cells, including proliferation, migration, and adhesion. Disturbance of these processes may cause vascular dysplasia.

Type II: ALK1 normally functions in the control of blood vessel development and repair. Most mutations create null alleles. The missense mutations lead to reduced ALK1 protein activity. ALK1 has a function in TGF-β signaling. Vascular dysplasia caused by haploinsufficiency of *ALK1* leads to telangiectases (only the endothelial wall is present) and arteriovenous malformations of the skin, mucosa, and viscera.

Type III: Unknown

Diagnosis

This syndrome may be diagnosed if three out of four criteria – epistaxis, telangiectases, visceral lesions, and family history – are present. With two criteria the syndrome is suspected, and with less than two criteria it is unlikely. Mutation analysis will confirm the diagnosis. Microscopy of the capillary pattern of the fingernail folds may be useful in diagnosis. Various vascular abnormalities of the capillaries should be looked for.

Types II and III show reduced incidence of pulmonary arteriovenous malformations. Type III is intermediate between types I and II. Vascular malformations in the liver might be a feature of type III.

Counseling issues

Penetrance of the disease is about 97%. Family history is negative in around 20% of patients. The *ENG* gene may be responsible for about 50% of inherited cases of all types. Nose bleeds can be controlled by hormone therapy (estrogen and progesterone), electro- or chemo-cauterization, laser treatment, closure of both nostrils, brachytherapy, vessel embolization, or dermoplasty. Pulmonary arteriovenous malformations can be controlled by embolotherapy or surgery. Almost all patients need oral iron supplementation. Bleeding is usually aggravated by pregnancy. In general, less than 10% of patients die of complications. Children of affected parents should be suspected of having the disorder – because of age-related penetrance, the disease may be less obvious than in adults. Early diagnosis will contribute to early detection of visceral involvement. Patients with pulmonary arteriovenous malformations or with a positive family history for these malformations (in cases where they have not been screened) should receive antibiotic prophylaxis before dental procedures or surgery, even if the malformations have already been surgically corrected. This will minimize the risk of developing septic emboli, which can cause a brain abscess.

5. Inherited Diseases: Miscellaneous

Velocardiofacial Syndrome

(also known as: VCF syndrome; Shprintzen syndrome)

MIM	192430
Clinical features	Submucous (10%–35% of patients) or overt (also 10%–35%) cleft palate, velopharyngeal inadequacy (30%–45%), speech and language impairment (70%–90%), hypernasality (45%–80%), learning disabilities, and mild intellectual impairment (50%). Cardiac features include ventricular septal defects (65%–75%), right aortic arch (35%–50%), tetralogy of Fallot (15%–20%), and an aberrant left subclavian artery (15%–20%). Typical facial features include a prominent nose with squared nasal root and narrow alar base, long face with vertical maxillary excess (85%), narrow palpebral fissures (35%–50%), retruded mandible (80%), long philtrum and thin upper lip, abundant scalp hair (50%), and microcephaly (40%–50%) (see **Figure 1**).
	Ophthalmologic abnormalities may be present, including tortuous retinal vessels, small optic discs, posterior embryotoxon, or bilateral cataracts (70%). Ear and hearing features include conductive hearing impairment due to frequent serous otitis media secondary to cleft palate or velopharyngeal inadequacy, SNHI (8%), and small auricles (40%–50%). Other features are hypotonia in infancy and childhood, scoliosis (10%), umbilical or inguinal hernia (25%), hypospadias (10%), renal abnormalities, hypocalcemia, immune disturbances, slender hands and fingers (60%), laryngotracheomalacia and glottic stenosis, microsomia (30%–40%), short stature (usually below the 10th percentile), and psychiatric disorders (10%–20%).
Age of onset	Congenital
Epidemiology	The prevalence of chromosome 22q11 deletions (see below) has been estimated at 1:4,000–5,000 (including both velocardiofacial syndrome and DiGeorge syndrome, next entry).
Inheritance	Autosomal dominant
Chromosomal location	22q11

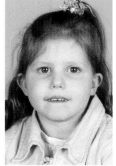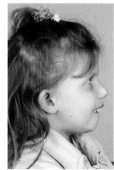

Figure 1. A 4-year-old girl with velocardiofacial syndrome. Note the prominent nose with a squared nasal root and narrow alar base, open mouth, and retruded mandible. The auricles are relatively small and posteriorly rotated, and there is a congenital short palate with velopharyngeal insufficiency.

Genes	Unknown
Mutational spectrum	There is a deletion in chromosome 22q11.
Effect of mutation	Unknown
Diagnosis	The clinical features suggest the diagnosis, which can be confirmed by chromosomal analysis using fluorescence *in situ* hybridization and microarray techniques. Significant phenotypic overlap exists with DiGeorge syndrome.
Counseling issues	Expression is variable. The deletion occurs *de novo* in approximately 85% of cases. Patients with a severe heart defect are usually diagnosed in the neonatal period; these defects need early surgical correction. Minor heart defects often go unnoticed until the age of 3–6 years. Palatal repair is necessary in cases of an overt cleft palate or submucous cleft. Ophthalmologic and nephro-urologic evaluation are required. Intelligence and psychosocial adjustment should be evaluated and parents advised as to education and school orientation. Speech–language and hearing should be assessed. Speech therapy is almost always necessary. The indication for pharyngoplasty can be assessed by nasopharyngoscopy and videofluoroscopy. Grommets are often needed in cases of longstanding otitis media with effusion. Hearing aids can be prescribed in cases of SNHI.

DiGeorge Syndrome

MIM	188400, 601362
Clinical features	Cardiac defects (95% of patients), including right aortic arch, interrupted aortic arch, ventricular septal defects, truncus arteriosus, patent ductus arteriosus, and tetralogy of Fallot. Hypoplasia or aplasia of the parathyroid glands produces hypoparathyroidism, causing severe hypocalcemia and seizures in early infancy. Hypoplasia or aplasia of the thymus gland cause T-cell immunodeficiency (and thus reduced defense against infection). Facial dysmorphisms (60%) include micrognathia; small, low-set, posteriorly angulated ears; anteverted nostrils; hypertelorism; and short philtrum. Additional features are mental deficiency, cleft palate, middle-ear, eye, and central nervous system abnormalities, esophageal atresia, choanal atresia, imperforate anus, and diaphragmatic hernia.
Age of onset	Congenital
Epidemiology	The prevalence of 22q11 deletions (see below) is estimated to be 1:4,000–5,000 (including velocardiofacial syndrome, previous entry). The prevalence of DiGeorge syndrome alone is unknown.
Inheritance	Most cases are sporadic. Some reports mention autosomal dominant and recessive inheritance.
Chromosomal location	22q11.2
Genes	Unknown
Mutational spectrum	There is a deletion in chromosome 22q11.2.
Effect of mutation	Some reports mention that deletion of 10p13–p14 can also cause the DiGeorge/velocardiofacial syndrome spectrum of malformations.
Diagnosis	Classic diagnostic criteria include two of the three major features: congenital heart defects, cellular immunodeficiency (by thymic hypoplasia or aplasia), and hypocalcemia (by parathyroid hypoplasia or aplasia). Chromosomal analysis with fluorescence *in situ* hybridization can confirm the diagnosis. There is considerable overlap between velocardiofacial syndrome (also [del]22q11) and DiGeorge syndrome.

Counseling	Expression is extremely variable. Hypocalcemia is treated in the neonatal period and cardiac defects are corrected surgically. Infections are frequent. Surgical repair of the cleft palate is performed. Speech therapy is almost always necessary.

Angioneurotic Edema

Angioneurotic edema usually occurs as an allergic response to exogenous factors, such as specific foods (eg, seafood), drugs, or temperature. This sporadic type of angioneurotic edema is present in up to 10% of the population. Angioneurotic edema may also be caused by angiotensin-converting enzyme (ACE) inhibitors via a similar, nonallergic mechanism. Therapy consists of adrenaline, steroid, and antihistamine administration.

In some cases, edema is the result of an intrinsic defect, which may be inherited or acquired. The more rare, acquired form can be caused by accelerated consumption of C1 esterase inhibitor (C1-INH). This may be a result of circulating antibodies that bind C1-INH, or of lymphoproliferative disease, other malignancies, or infection. In the autoantibody-mediated type, symptoms can be controlled with infusion of a very high dose of C1-INH concentrate.

In all the above-mentioned types, ACE inhibitors should not be prescribed.

Angioneurotic Edema, Hereditary

MIM	106100
Clinical features	Episodes of subcutaneous skin swelling and submucosal edema of the upper airway and gastrointestinal tract. Most patients have at least one attack a month. Upper airway obstruction is usually caused by laryngeal edema; however, some patients show pulmonary edema. Laryngeal symptoms are dyspnea, dysphagia, and voice changes. The episodes of edema may last 2–5 days and can be life threatening. Trauma (eg, intubation, operations), anxiety, stress, menstruation, infections, oral contraceptives, and angiotensin-converting enzyme (ACE) inhibitors may precipitate or aggravate edema. Gastrointestinal involvement includes abdominal pain, nausea, diarrhea, ascites, and vomiting; occasionally, these are the only symptoms. Low levels of C4 can cause systemic lupus erythematosus, glomerulonephritis, and vasculitis. Women with this disease have a high prevalence of polycystic ovaries.

Age of onset	Presentation is often in the second decade with mild symptoms, which become worse after puberty.
Epidemiology	Approximately 1:10,000–50,000
Inheritance	Autosomal dominant
Chromosomal location	11q11–q13.1
Genes	*C1-INH* (complement 1 esterase inhibitor)
Mutational spectrum	The majority of mutations are Alu-mediated deletions and duplications. Nonsense mutations, small insertions, and deletions are also seen.
Effect of mutation	Two types of hereditary angioneurotic edema are differentiated. Type I patients have either a deletion of the gene or a mutation leading to a truncated C1-INH protein; C1-INH levels are 5%–30% of normal. This type represents about 85% of patients. Mutations in type II patients lead to amino-acid substitutions; C1-INH serum levels are normal or even elevated, but the protein is functionally deficient. Insufficiency (dysfunction or deficiency) of activated C1-INH due to defective hepatic synthesis causes edema via an incompletely understood mechanism.
Diagnosis	Types I and II cannot be distinguished clinically. All patients show C1-INH insufficiency (functional test) and often have low C2 and C4 levels. The C1q level is normal, which is not the case in acquired forms.
Counseling issues	About 20%–25% of mutations are *de novo*, and family history is therefore absent. Therapy is determined by indication.
	(1) Long-term prophylaxis. Danazol (impeded androgen) administration stimulates the normal allele to produce more C1-INH. Adverse effects include weight gain, menstrual irregularities, arterial hypertension, and (possibly) hepatocellular adenoma. Long-term administration of tranexamic acid is less effective. Its main adverse effect is an increased risk of thromboembolism. Administration of C1-INH is effective in preventing attacks; however, it is expensive and the concentrate may transmit disease.
	(2) Short-term prophylaxis before surgery. This consists of administration of either C1-INH or danazol.

(3) Treatment of acute attacks. In most patients, laryngeal symptoms start to resolve within 30–60 minutes of infusion of C1-INH concentrate. In untreated patients, the mean duration of laryngeal edema episodes is 3–5 days, compared with 5–24 hours in treated patients. In mild cases of laryngeal edema, hospitalization – possibly with oxygen therapy – may be sufficient.

In pregnant women, delivery may be difficult because of perineal edema, and abdominal pain may mask other obstetric signs. Purified C1-INH concentrate may be infused as a prophylaxis in these cases.

The frequency of abdominal symptoms is significantly higher in patients who suffer from *Helicobacter pylori* infection. Bacterial eradication therapy can lead to a dramatic decrease in symptoms. This suggests that patients should be screened for *H. pylori* infection. In general, patients have a higher risk of developing autoimmune diseases (up to about 12%; 2% suffer from lupus erythematosus).

ACE inhibitors and oral contraceptives should not be prescribed to patients with hereditary angioneurotic edema. Patients' relatives should be tested and receive genetic counseling. Some of those affected will never develop clinical symptoms.

Notes   Autosomal recessive inheritance has been reported in one family, causing a low C1-INH level and severe clinical symptoms. Heterozygotes in this family were free from attacks. An X-linked dominant form (MIM 300268), which only occurs in women, has also been described. All features are the same as for types I and II, described above; however, a normal serum C4 concentration and C1-INH concentration and function are found.

Laryngeal Paralysis, Familial

Vocal fold paralysis is the second most common cause of congenital stridor (after laryngomalacia). Vocal fold paralysis may result in hoarseness, dysphagia, and recurrent aspiration pneumonia. Most vocal fold paralyses have an idiopathic, neurologic, or iatrogenic origin; few patients have a positive family history.

(a) Abductor Paralysis

MIM	Not listed
Clinical features	Variable laryngeal abductor paralysis. Severe obstruction often makes a tracheotomy necessary in the neonatal period. Affected individuals show normal laryngeal adduction. In the first years of life, vocal fold movement might show some recovery (there is complete recovery in some cases) and patients with tracheotomies may be decannulated.
Age of onset	Congenital
Epidemiology	One family has been described.
Inheritance	Autosomal dominant
Chromosomal location	6q16
Genes	Unknown
Mutational spectrum	Unknown
Effect of mutation	The nucleus ambiguus is probably affected through a developmental delay.
Diagnosis	Family history and clinical features may suggest the diagnosis. This can be confirmed by linkage analysis.
Counseling issues	Emergency intubation and/or tracheotomy might be necessary shortly after birth. Expression of the syndrome is variable, with nonpenetrance at its extreme.

(b) Plott Syndrome

MIM	308850
Clinical features	Permanent laryngeal abductor paralysis, often with stridor and mental retardation.
Age of onset	Congenital

Epidemiology	Approximately five families have been described.
Inheritance	X-linked recessive inheritance has been suggested.
Chromosomal location	Unknown
Genes	Unknown
Mutational spectrum	Unknown
Effect of mutation	Dysgenesis of the nucleus ambiguus has been suggested.
Diagnosis	Family history and clinical features can lead to a diagnosis.
Counseling issues	Emergency intubation and/or tracheotomy may be necessary shortly after birth. Mental retardation is possibly due to brain damage resulting from respiratory distress.

(c) Gerhardt Syndrome

MIM	150260
Clinical features	Laryngeal abductor paralysis, often with stridor. Some families show subtle central neurologic abnormalities, such as abnormalities in auditory brainstem response (ABR) testing or pharyngoesophageal manometric studies, and/or swallowing difficulties after birth. Mental retardation is present in some patients. Vocal fold movement (abduction) may show some recovery in the first years of life, and patients with tracheotomies may be decannulated.
Age of onset	Congenital
Epidemiology	Five to 10 families have been described.
Inheritance	Autosomal dominant
Chromosomal location	Unknown
Genes	Unknown

Mutational spectrum Unknown

Effect of mutation The nucleus ambiguus is probably affected by delayed development.

Diagnosis Family history and clinical features suggest the diagnosis.

Counseling issues Emergency intubation and/or tracheotomy may be necessary shortly after birth. Expression and penetrance of the syndrome are variable. Occasionally, inherited laryngeal abductor paralysis is found as part of a broader neurologic disease. In one family, bilateral recurrent laryngeal nerve paralysis in combination with bilateral ptosis was described as an autosomal dominant disorder.

(d) Adductor Paralysis

MIM 150270

Clinical features Congenital hoarseness with some progression on aging. There is no mental retardation.

Age of onset Congenital

Epidemiology One family has been described.

Inheritance Autosomal dominant

Chromosomal location 6p21.3–p21.2

Genes Unknown

Mutational spectrum Unknown

Effect of mutation Unknown

Diagnosis Familial congenital hoarseness indicates the diagnosis, which can be confirmed by linkage analysis.

Counseling issues None. Operative intervention was not performed in the one family so far described.

Opitz Syndrome, Types I and II

(also known as: BBB syndrome; G syndrome; BBBG syndrome;
opitz-oculo-genito-laryngeal syndrome)

MIM	300000 (type I), 145410 (type II)
Clinical features	In some literature, types I and II are differentiated; however, no firm phenotypic differentiation has yet been established. Opitz syndrome is characterized by midline defects, including hypertelorism, hypospadias, and laryngotracheoesophageal defects (see **Figure 2**). Hypertelorism or dystopia canthorum (together, 90% of patients), up/downslanting palpebral fissures, epicanthic folds, and strabismus may be found. Some patients have relative entropion. The nasal bridge may be elevated or flattened, nostrils are anteverted (type I patients only), and the philtrum is flat and unapparent. Micrognathia and cleft lip/palate are found in about 30% of patients. A broad or bifid uvula, ankyloglossia, bifid tongue, short lingual frenulum, supernumerary teeth, and malocclusion might be present. A posterior pharyngeal cleft can be observed (type I only). The pinnae may be rotated posteriorly; sporadically, they are an abnormal shape. The disease can present with cranial asymmetry (20%), brachycephaly, prominent forehead, open fontanels, and prominence of occipital and parietal eminences.

Laryngotracheoesophageal defects are found in about 40% of patients. These patients can present with stridor, a hoarse cry, choking, coughing, and cyanosis. This may be followed by respiratory distress and pneumonia. Abnormal esophageal motility is found in about 70% of patients. Approximately 40% show renal anomalies, ureteral stenosis, and inguinal hernia. Ectopic and imperforate anus have been reported. Hypospadias, which is found in all Opitz males, has a marked variability in expression. Congenital heart defects are found in 25% of patients. Cerebral imaging can reveal various anomalies. Mild to moderate mental retardation is found in 65% of patients. It is noteworthy that many of the above-mentioned features concern midline defects. Hypertelorism might be the only feature in affected females, and some female carriers even appear to have a normal phenotype. Affected females have normal genitalia.

Major anomalies are more often found in male type I patients than in type II patients. The phenotype of both types is more complex and more severe in male than in female patients.

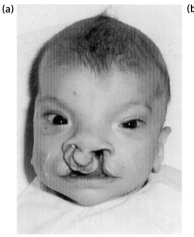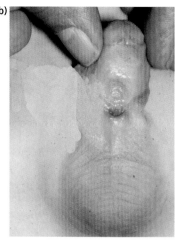

Figure 2. (a) A baby with Opitz syndrome showing midline defects – in this case hypertelorism, greater than expected (on the basis of hypertelorism alone) lateral displacement of the inner canthus of each eye, and a cleft lip and palate. Other features are upslanting palpebral fissures and a flat nasal bridge. **(b)** The same patient showing a hypospadias.

Age of onset	Congenital
Epidemiology	Several families have been described.
Inheritance	X-linked (dominant; type I); autosomal dominant, variable expression (type II)
Chromosomal location	Xp22 (type I), 22q11.2 (type II)
Genes	*MID1* (midin; type I); unknown for type II
Mutational spectrum	In type I, gross insertions, deletions, and duplications are found, as well as small deletions and insertions, missense mutations, and nonsense mutations. In addition, an inversion breakpoint that disrupts the gene has been described. Chromosome 22q11.2 deletion has been found in type II.
Effect of mutation	In type I, the majority of the mutations lead to protein truncation. MID1 is a microtubule-associated protein with E3 ubiquitin ligase activity. There is experimental evidence that normal MID1 protein targets PP2Ac for degradation. Accumulation of PP2Ac leads to increased

dephosphorylation of microtubule-associated proteins and disturbed microtubule dynamics. This might affect apoptosis and cell migration, which is important for ventral midline development. The effect of (del)22q11.2 in type II is unknown.

Diagnosis

Clinical features suggest the diagnosis, which can be confirmed by mutation analysis in type I. Radiologic swallowing studies will reveal laryngotracheoesophageal defects. In some cases, Opitz type I can be diagnosed by prenatal ultrasound. In type II, some phenotypic overlap may be found with velocardiofacial (p. 154) and DiGeorge (p. 156) syndromes, which are also associated with chromosome 22q11 deletions.

Counseling issues

Patients with laryngotracheoesophageal defects have a high mortality rate unless the defect is repaired. Monozygotic twinning may be an expression of the syndrome. In patients who wish for cosmetic nose surgery, the procedure can be complicated by agenesis of the nasal cartilage. Auricular cartilage or homograft cartilage can be used in reconstruction. Disease expression is variable.

Paraganglioma

(also known as: PGL; nonchromaffin paraganglioma; glomus tumors; chemodectomas; carotid body tumors; glomus jugulare tumors; vagal body tumors)

Paraganglioma arise from paraganglia: minute bodies of neuroendocrine tissue that histologically resemble adrenal medulla and are able to produce and secrete catecholamines (dopamine, norepinephrine, and epinephrine). In contrast to paraganglions in the lower parts of the body, which are closely related to the sympathetic system, paraganglions in the head and neck are parasympathetically innervated. The carotid body (situated at the carotid bifurcation) serves as a chemoreceptor organ and is connected to the ventilation regulatory system through Hering's nerve (a branch of the glossopharyngeal nerve). Carotid body tumors account for 60%–80% of paragangliomas of the head and neck. Tumors of the tympanic body on the promontory in the middle ear and the jugular body at the jugular bulb are often clinically indistinguishable and together account for 25%–50% of clinically affected patients. The vagal body at the nodose ganglion of the vagal nerve is affected in 5%–20% of patients. Finally, rare paragangliomas have been described in the laryngeal body of

Type	Description
I	Situated in the carotid bifurcation
II	The carotid arteries are partially surrounded
III	The carotid arteries and hypoglossal nerve are completely surrounded

Table 1. Shamblin's classification of carotid body tumors.

Type	Description
A	The tumor is confined to the tympanum
B	The tumor arises from the hypotympanum and the cortical bone over the jugular bulb is intact
C	The bone over the jugular bulb is eroded
D	The tumor has an intracranial extension

Table 2. Fisch's classification of jugulotympanic tumors.

the larynx (via the superior and inferior laryngeal nerves), the cardioaortic bodies of the supracardiac region, and the ciliary body of the orbit.

Shamblin's classification of carotid body tumors differentiates three types (see **Table 1**). Vagal body tumors do not have an official classification. Jugulotympanic tumors are classified according to Fisch (see **Table 2**).

Paragangliomas usually have a very slow growth pattern and can advance to a considerable size without causing any major symptoms. Aggressive invasive growth is observed in a minority of cases. Head and neck paragangliomas may occasionally metastasize (about 1%–4%) to regional lymph nodes or hematogenous. The natural cause of metastases of paragangliomas is similarly variable, as is the behavior of primary tumors.

Carotid body tumors usually present as a large, possibly tender mass in the neck. Jugular bulb and tympanic body paragangliomas often lead to conductive hearing impairment and pulsatile tinnitus. Otorrhea resulting from secondary infection and spontaneous bleeding has been described. These symptoms generally lead to an earlier diagnosis than cases of carotid body tumors. Jugular paragangliomas, however, can also present with lower cranial nerve palsy due to involvement of the jugular foramen. Vertigo and SNHI can indicate labyrinthine involvement. Jugulotympanic tumors can also lead to facial nerve palsy and intracranial complications due to intracranial extension. Vagal body tumors present with a pharyngeal or neck mass. Hoarseness due to vocal cord paralysis is sometimes the first symptom.

Treatment of head and neck paragangliomas is a moot point. Surgery of Shamblin type I and II carotid body tumors and Fisch type A jugulotympanic tumors can be performed without major complications. In all larger tumors, there is a high risk of lower cranial nerve damage.

The estimated incidence for all types is 1:400,000 persons each year, but this may be an underestimation because of asymptomatic and untreated disease, which is not included in this figure. The estimated prevalence for both inherited and sporadic cases is 1:30,000.

Hereditary paragangliomas have been recognized since the beginning of the 20th century. Currently, four different genetic types of paraganglioma are recognized: PGL1, PGL2, PGL3, and "carotid body tumors and multiple extra-adrenal pheochromocytomas".

On the basis of family history, the prevalence of hereditary cases among the total of head and neck paragangliomas has been estimated at 10%. DNA studies have shown that a considerable number of solitary cases are caused by inherited germline mutations. Due to founder effects, the prevalence of hereditary cases can exceed 50% locally. Hereditary cases tend to develop multiple paragangliomas. Bilateral carotid body tumors exist in about 30% of patients in families at risk. In approximately 50%, other paragangliomas occur, including pheochromocytomas.

Evaluation of serum and urine catecholamine levels can lead to the identification of a concomitant pheochromocytoma or a pheochromocytoma-like syndrome. The latter may result from a head and neck paraganglioma (1% of head and neck paragangliomas) that excretes large amounts of catecholamines. These patients often suffer from headaches, palpitations, flushing, and labile hypertension. Preoperatively, these patients require α-adrenergic blockage.

(a) Paraganglioma, Type I

(also known as: PGL1)

MIM	168000
Clinical features	The most common of inherited paragangliomas, affecting (in decreasing order) the carotid body, jugulotympanic bodies, and vagal body. Includes pheochromocytoma in at least 5% of cases. See the introduction for further clinical features.

Age of onset	Usually in the third or fourth decade, but occasionally in teenagers. A delay of several years is common between the onset of symptoms and diagnosis.
Inheritance	Autosomal dominant, with incomplete, age-dependant penetrance. The penetrance is zero after female transmission. This inheritance mechanism may be caused by genomic imprinting. The *SDHD* gene itself (see below), however, is not imprinted in the tissues that are tested. Sporadic cases are frequent.
Chromosomal location	11q23
Genes	*SDHD* (succinate dehydrogenase complex, subunit D)
Mutational spectrum	Missense, nonsense, splice-site, and frame-shift mutations have been found.
Effect of mutation	The mutations result in defective succinate–ubiquinone oxidoreductase (mitochondrial respiratory chain complex II). It has been suggested that the defective oxygen-sensing system gives rise to cellular proliferation.
Diagnosis	A positive family history or multiple paragangliomas are suggestive. Diagnosis can be reached by mutation analysis. Magnetic resonance imaging (see **Figure 3**) of the entire head and neck region or octreotide scintigraphy should be performed to exclude occult secondary paragangliomas. Serum and urine catecholamine levels must be evaluated to exclude pheochromocytomas. Ultrasound-guided fine-needle aspiration cytology may be helpful if the diagnosis remains unclear.
Counseling issues	Mutations in the *SDHD* gene are the leading cause of head and neck paragangliomas. The inheritance pattern is unique due to its complete suppression of penetrance after female transmission. The disease therefore skips generations in the maternal line and patients may not have a family history of paraganglioma. The penetrance of the disease is age-related and not all patients develop symptoms during their lifetime. Up to 40% of the anamnestically sporadic cases are actually found to have an inherited tumor after genetic analysis. Presymptomatic MRI scans can be made on a regular basis in patients carrying an *SDHD* gene mutation. Surgical intervention is not always necessary, but early diagnosis keeps all the treatment options

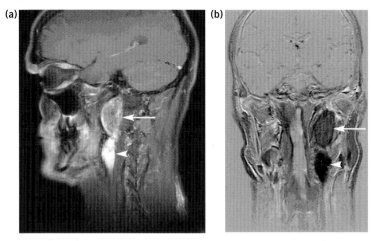

Figure 3. A 40-year-old male patient suffering from multiple inherited paragangliomata on the left. (**a**) A T1-weighted magnetic resonance image with fat saturation and (**b**) a T1-weighted subtraction image, both after intravenous gadolinium. The more cranially located lesion is a vagal body tumor (arrow) and the more caudally situated lesion represents a carotid body tumor (arrowhead). In this case, a "wait and scan" policy is recommended.

open. In cases where no intervention is chosen, the natural course of the disease can at least be monitored.

(b) Paraganglioma, Type II

(also known as: PGL2)

MIM	601650
Clinical features	The *PGL2* locus has been linked to hereditary paragangliomas in a single large multibranch family from The Netherlands. See the introduction for further clinical features.
Age of onset	On average, 37 years (range: 16–80 years). This is earlier than for PGL1. The average delay between onset and diagnosis is 11 years.
Inheritance	Similar to PGL1. Autosomal dominant, with incomplete, age-dependant penetrance. The penetrance is zero after female transmission.
Chromosomal location	11q13.1

Genes	Unknown
Mutational spectrum	Unknown
Effect of mutation	Unknown
Diagnosis	See PGL1. Intrafamilial diagnosis can be made by linkage analysis.
Counseling issues	See PGL1. However, PGL2 seems to reach 100% penetrance by the age of 45 years.

(c) Paraganglioma, Type III

(also known as: PGL3)

MIM	605373
Clinical features	Like PGL2, PGL3 is limited to a single (German) family, including five patients with head and neck paragangliomas. A solitary case with PGL3 has also been identified. This patient had a catecholamine-secreting tumor and regional lymph node metastasis.
Age of onset	Adulthood
Inheritance	Autosomal dominant. Only maternal transmission has been described.
Chromosomal location	1q21
Genes	*SDHC* (succinate dehydrogenase complex, subunit C)
Mutational spectrum	A nucleotide change that destroys the start codon for protein synthesis and splice-site mutations have been described.
Effect of mutation	Mutations result in a defect in succinate–ubiquinone oxidoreductase (mitochondrial respiratory chain complex II).
Diagnosis	See PGL1
Counseling issues	See PGL1. However, full penetrance is reported and no imprinting is found.

(d) Carotid Body Tumors and Multiple Extra-adrenal Pheochromocytomas

MIM	115310
Clinical features	This disease is characterized by adrenal paragangliomas (pheochromocytomas), functional retroperitoneal extra-adrenal paragangliomas, and head and neck paragangliomas.
Age of onset	Adulthood
Inheritance	Autosomal dominant
Chromosomal location	1p36.1–p35
Genes	*SDHB* (succinate dehydrogenase complex subunit B)
Mutational spectrum	Nucleotide changes that cause amino-acid substitutions or premature protein truncation have been found; also, small insertions and deletions leading to protein truncation.
Effect of mutation	Mutations result in a defect in succinate–ubiquinone oxidoreductase (mitochondrial respiratory chain complex II).
Diagnosis	See PGL1. In addition, this syndrome should be suspected in patients/families with extra-adrenal pheochromocytoma without evidence of cervical paragangliomas.
Counseling issues	No parent-of-origin effects are known. Patients with a mutation in this gene should be tested for the presence of pheochromocytomas. Although the phenotypes of von Hippel Lindau (VHL) disease and multiple endocrine neoplasia (MEN2) are more extensive, patients with hereditary pheochromocytoma should also be tested for mutations in the *VHL* and *RET* genes.

Primary Ciliary Dyskinesis

(also known as: dyskinetic cilia syndrome; immotile cilia syndrome)

(a) Primary Ciliary Dyskinesis with Situs Inversus

(also known as: Kartagener syndrome; Siewert syndrome)

MIM	244400
Clinical features	Situs inversus (chest and abdomen), bronchitis leading to bronchiectasis, sinusitis, nasal polyps, and infertility result from impaired ciliary function. Recurrent pneumonia and chronic coughing – resulting from reduced mucociliary clearance – can occur. Associated cardiac anomalies may be seen in combination with dextrocardia. If dextrocardia is absent, there is isolated primary ciliary dyskinesis (see below). Some patients suffer from anosmia, rhinitis, headaches (ciliated brain tissue), and corneal abnormalities (monociliated tissue). The temporal bone may be less aerated and show chronic otitis media with conductive hearing impairment. The frontal sinus can be absent. Women have lowered fertility and men are usually infertile. However, *in vitro* fertilization and intracytoplasmic sperm injection are possible with sperm from these males. Some affected males show normal sperm motility.
Age of onset	Congenital
Epidemiology	1:30,000–60,000
Inheritance	Autosomal recessive
Chromosomal location	19q13.3–qter, 9p21–p13 (*DNAI1*), 5p15–p14 (DNAH5)
Genes	*DNAI1* (dynein axonemal intermediate chain 1), *DNAH5* (dynein axonemal heavy chain 5)
Mutational spectrum	Splice-site mutations, missense mutations, and small insertions and deletions are found in both *DNAI1* and *DNAH5*. Missense mutations are detected in conserved amino acids. The majority of mutations in both *DNAI1* and *DNAH5* are predicted to lead to protein truncation. In

DNAI1, 219+3insT is the most common mutation. Linkage to chromosome 19 has been described in Arabic families.

Effect of mutation Cilia in the respiratory tract and sperm show absence/abnormality of dynein arms of the microtubules, which form the skeleton of the cilia. These dynein arms form temporary bridges between ciliary filaments, resulting in movement of cilia and sperm tails. Their absence/deformity results in immobility or abnormal ciliary beating. In addition, radial spokes are abnormal and the central pair of microtubules might be absent. Normal movement of cilia in embryonic tissues is believed to cause normal visceral asymmetry; abnormal movement causes random lateralization (ie, the organs are randomly lateralized on the left or right side).

Diagnosis This disease should be suspected in the presence of chronic otitis media, sinusitis, and bronchitis in a patient with situs inversus, which may be accompanied by infertility. Confirmation might be obtained by electron microscopic examination of a tracheal or nasal biopsy. Biopsies should not be taken after an acute respiratory tract infection. Chest and abdominal x-rays are used to diagnose situs inversus and/or bronchiectasis. Sperm dysmotility might be observed. In European and North American families, mutations are found in about 40% of cases.

Counseling issues Chest physiotherapy should be given. Drug therapy may consist of bronchodilators and antibiotics for pulmonary infections, acute otitis media, and sinusitis. An otorhinolaryngologist might be consulted regarding chronic sinusitis, rhinitis, nasal polyps, and otitis media with effusion. If sinusitis persists, a nebulizer can be prescribed for the nasal deposition of antibiotics, or an antrostomy under the inferior nasal concha is an option. Insertion of a ventilation tube (for chronic otitis media and conductive hearing impairment) may be postponed since chronic otorrhea and a permanent perforation are frequently seen. Successful tympanoplasties have been reported in children with perforations, chronic otorrhea, and/or hearing impairment. Following this operation, otorrhea may disappear and hearing improve. In cases of otitis media with effusion and a conductive hearing impairment, the temporary use of a hearing aid should be considered. Heart and/or lung transplantations have been performed in some patients. The overall morbidity is significant, although many patients live a normal and active life. Life expectancy is unknown.

(b) Primary Ciliary Dyskinesis without Situs Inversus

MIM	242650
	Situs inversus is absent in these families, but further features and management are the same as described in the previous entry. Families without the presence of situs inversus have been described, however, most families contain both members with the Kartagener syndrome and with primary ciliary dyskinesis without situs inversus. X-linked and autosomal dominant inherited forms probably exist.

White Sponge Nevus

(also known as: spongiosus albus mucosae)

MIM	193900
Clinical features	In this benign disorder, the oral mucosa is bilaterally spongy-fold and has a white opalescent tint (oral leukokeratosis). Buccal mucosa is most often involved (see **Figure 4**), followed by labial and gingival mucosa, the floor of the mouth, and the lateral borders of the tongue. Mucosa of the esophagus, vagina, rectum, and nasal cavity might also be involved. Extraoral lesions are relatively uncommon in the absence of oral lesions. Extramucosal lesions do not appear. The lesions are usually painless and asymptomatic, although pruritis, burning, and altered texture and dryness of the mouth can occur. Lesions usually remain unchanged over time. However, periods of remission and exacerbation may be seen.
Age of onset	Any age from birth onwards.
Epidemiology	Rare
Inheritance	Autosomal dominant
Chromosomal location	17q21–q22, 12q13
Genes	*KRT4* (keratin 4), *KRT13* (keratin 13)
Mutational spectrum	A 3-bp deletion and 3-bp insertion have been described in *KRT4*, leading to a deletion (N159) or insertion of one amino acid (T154),

Figure 4. The buccal mucosa of a patient suffering from white sponge nevus. Figure courtesy of I van der Waal.

respectively. Two missense mutations, N112S and L119P, have been detected in *KRT13*.

Effect of mutation   All three mutations affect the helix initiation peptide in the 1A region of keratins 4 and 13 (this is a conserved region in keratin proteins). The nonepidermal keratins K4 and K13 are coexpressed in suprabasal keratinocytes of mucous membranes, where K4 and K13 heterodimerize. The mutations are predicted to impair keratin filament assembly and/or integrity. This results in thickening of the epithelium, degeneration of the suprabasal cells, and clumps or aggregates of keratin filaments.

Diagnosis   Clinical diagnosis is confirmed by biopsy. Mutation analysis leads to a definitive diagnosis. This disease should be differentiated from hereditary benign intraepithelial dyskeratosis (MIM 127600). The latter also shows white lesions of the oral mucosa and is autosomal dominantly inherited; however, vaginal and anal lesions are not present, and conjunctival involvement is seen.

Counseling issues   Disease penetrance is about 87%. Sporadic cases occur. Usually, no treatment is indicated. In symptomatic cases, topical treatment with tretinoin (acid form of vitamin A) and/or topical antibiotics (eg, tetracycline, penicillin) should be considered. A response to different oral antibiotics (amoxycillin, tertracycline) has also been observed, with remission continued under low-dose maintenance treatment. Mouth rinses for oral lesions and antipruritic creams for anogenital lesions can be helpful.

6. Abbreviations

AAID	acute aminoglycoside-induced deafness
ABR	auditory brainstem response
ACE	angiotensin-converting enzyme
ALK	activin receptor-like kinase
ATP	adenosine triphosphate
BER	brainstem-evoked responses
BO	branchio-oto
BOR	branchio-oto-renal
BR	branchio-renal
C1-INH	C1 esterase inhibitor
CHARGE	coloboma of the eye (C), heart anomaly (H), atresia choanae (A), retardation of growth and/or development (R), genitourinary anomalies (G), ear anomaly, hearing impairment, or deafness (E)
CI	cochlear implantation
CNS	central nervous system
COX	cytochrome C oxidase
CPEO	chronic progressive external ophthalmoplegia
CSF	cerebrospinal fluid
CT	computed tomography
DDP	deafness/dystonia peptide
DIDMOAD	diabetes insipidus, diabetes mellitus, optic atrophy, and deafness
DNA	deoxyribonucleic acid
DPP	dentin phosphoprotein
DSP	dentin sialoprotein
DSPP	dentin sialophosphoprotein
ENG	electronystagmogram
ENT	ear, nose, and throat
ERG	electroretinogram
ESRD	end-stage renal disease
EVA	enlarged vestibular aqueduct
FGFR	fibroblast growth factor receptor
GnRH	gonadotropin-releasing hormone
JLNS	Jervell and Lange–Nielsen syndrome
KAL	Kallmann syndrome
K4	keratin 4
KSS	Kearns–Sayre syndrome

LHON	Leber's hereditary optic neuropathy
MCPP	metacarpophalangeal pattern profile
MELAS	mitochondrial encephalomyopathy, lactic acidosis, stroke-like episodes
MEN	multiple endocrine neoplasia
MERRF	myoclonic epilepsy associated with ragged-red fibers
MID	midin
MIM	Mendelian Inheritance in Man
MNGIE	myoneurogastrointestinal encephalopathy
MRI	magnetic resonance imaging
mRNA	messenger RNA
mtDNA	mitochondrial DNA
NARP	neuropathy, ataxia, and retinitis pigmentosa
ND	NADH dehydrogenase
NF	neurofibromatosis
NMR	nuclear magnetic resonance
NS	Noonan syndrome
OSMED	otospondylomegaepiphysial dysplasia
OXPHOS	oxidative phosphorylation
PET	positron emission tomography
PGL	paraganglioma
PTA	pure tone average
RNA	ribonucleic acid
rRNA	ribosomal RNA
SH	Src homology region
SNHI	sensorineural hearing impairment
TGF-β	transforming growth factor-β
TIMM	translocase of inner mitochondrial membrane
tRNA	transfer RNA
TSH	thyroid-stimulating hormone
USH	Usher syndrome
VCF	velocardiofacial
WS	Waardenburg syndrome

7. Glossary

A

Adenine (A)   One of the bases making up **DNA** and **RNA** (pairs with **thymine** in DNA and **uracil** in RNA).

Agarose gel   See **electrophoresis.**
electrophoresis

Allele   One of two or more alternative forms of a **gene** at a given location (**locus**). A single allele for each locus is inherited separately from each parent. In normal human beings there are two alleles for each locus (**diploidy**). If the two alleles are identical, the individual is said to be **homozygous** for that allele; if different, the individual is **heterozygous**.

For example, the normal **DNA** sequence at **codon** 6 in the beta-globin gene is GAG (coding for glutamic acid), whereas in sickle cell disease the sequence is GTG (coding for valine). An individual is said to be heterozygous for the glutamic acid → valine **mutation** if he/she possesses one normal (GAG) and one mutated (GTG) allele. Such individuals are **carriers** of the sickle cell gene and do not manifest classical sickle cell disease (which is **autosomal recessive**).

Allelic heterogeneity   Similar/identical **phenotypes** caused by different **mutations** within a **gene**. For example, many different mutations in the same gene are now known to be associated with Marfan's syndrome (*FBN1* gene at 15q21.1).

Amniocentesis   Withdrawal of amniotic fluid, usually carried out during the second trimester, for the purpose of prenatal diagnosis.

Amplification   The production of increased numbers of a **DNA** sequence.

1. *In vitro*
In the early days of recombinant DNA techniques, the only way to amplify a sequence of interest (so that large amounts were available for detailed study) was to **clone** the fragment in a vector (**plasmid** or phage) and transform bacteria with the recombinant vector. The transformation technique generally results in the "acceptance" of a single vector molecule by each bacterial cell. The vector is able to exist autonomously within the bacterial cell, sometimes at very high copy numbers (eg, 500 vector copies per cell). Growth of the bacteria containing the vector, coupled

with a method to recover the vector sequence from the bacterial culture, allows for almost unlimited production of a sequence of interest. Cloning and bacterial propagation are still used for applications requiring either large quantities of material or else exceptionally pure material.

However, the advent of the **polymerase chain reaction** (PCR) has meant that amplification of desired DNA sequences can now be performed more rapidly than was the case with cloning (a few hours cf. days), and it is now routine to amplify DNA sequences 10 million-fold.

2. *In vivo*

Amplification may also refer to an increase in the number of DNA sequences within the genome. For example, the genomes of many tumors are now known to contain regions that have been amplified many fold compared to their nontumor counterparts (ie, a sequence or region of DNA that normally occurs once at a particular chromosomal location may be present in hundreds of copies in some tumors). It is believed that many such regions harbor **oncogenes**, which, when present in high copy number, predispose to development of the malignant **phenotype**.

Aneuploid

Possessing an incorrect number (abnormal complement) of **chromosomes**. The normal human complement is 46 chromosomes, any cell that deviates from this number is said to be aneuploid.

Aneuploidy

The chromosomal condition of a cell or organism with an incorrect number of **chromosomes**. Individuals with Down's syndrome are described as having aneuploidy, because they possess an extra copy of chromosome 21 (**trisomy** 21), making a total of 47 chromosomes.

Anticipation

A general phenomenon that refers to the observation of an increase in severity, and/or decrease in age of onset, of a condition in successive generations of a family (see **Figure 1**). Anticipation is now known, in many cases, to result directly from the presence of a **dynamic mutation** in a family. In the absence of a dynamic mutation, anticipation may be explained by "**ascertainment bias**". Thus, before the first dynamic mutations were described (in Fragile X and myotonic dystrophy), it was believed that ascertainment bias was the complete explanation for anticipation. There are two main reasons for ascertainment bias:

1. Identical **mutations** in different individuals often result in variable expressions of the associated **phenotype**. Thus, individuals within a

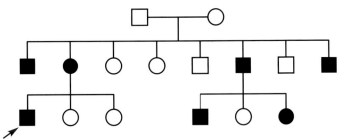

Figure 1. Autosomal dominant inheritance with **anticipation**. In many disorders that exhibit anticipation, the age of onset decreases in subsequent generations. It may happen that the transmitting parent (grandparent in this case) is unaffected at the time of presentation of the **proband** (see arrow). A good example is Huntington's disease, caused by the expansion of a CAG repeat in the coding region of the huntingtin gene. Note that this **pedigree** would also be consistent with either gonadal **mosaicism** or reduced **penetrance** (in the **carrier** grandparent).

family, all of whom harbor an identical mutation, may have variation in the severity of their condition.

2. Individuals with a severe phenotype are more likely to present to the medical profession. Moreover, such individuals are more likely to fail to reproduce (ie, they are genetic lethals), often for social, rather than direct physical reasons.

For both reasons, it is much more likely that a mildly affected parent will be ascertained with a severely affected child, than the reverse. Therefore, the severity of a condition appears to increase through generations.

Anticodon	The 3-base sequence on a **transfer RNA** (tRNA) molecule that is complementary to the 3-base **codon** of a **messenger RNA** (mRNA) molecule.
Ascertainment bias	See **anticipation**.
Autosomal disorder	A disorder associated with a **mutation** in an autosomal **gene**.
Autosomal dominant (AD) inheritance	An **autosomal disorder** in which the **phenotype** is expressed in the **heterozygous** state. These disorders are not sex-specific. Fifty percent of offspring (when only one parent is affected) will usually manifest the disorder (see **Figure 2**). Marfan syndrome is a good example of an AD disorder; affected individuals possess one wild-type (normal) and one mutated **allele** at the *FBN1* **gene**.

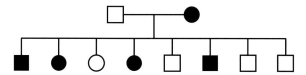

Figure 2. Autosomal dominant (AD) inheritance.

Autosomal recessive (AR) inheritance
An **autosomal disorder** in which the **phenotype** is manifest in the **homozygous** state. This pattern of inheritance is not sex-specific and is difficult to trace through generations because both parents must contribute the abnormal **gene**, but may not necessarily display the disorder. The children of two **heterozygous** AR parents have a 25% chance of manifesting the disorder (see **Figure 3**). Cystic fibrosis (CF) is a good example of an AR disorder; affected individuals possess two **mutations**, one at each **allele**.

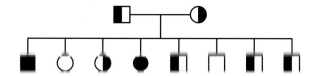

Figure 3. Autosomal recessive (AR) inheritance.

Autosome
Any **chromosome**, other than the **sex chromosomes** (X or Y), that occurs in pairs in **diploid** cells.

B

Barr body
An inactive **X chromosome**, visible in the **somatic cells** of individuals with more than one X chromosome (ie, all normal females and all males with Klinefelter's syndrome). For individuals with nX chromosomes, n−1 Barr bodies are seen. The presence of a Barr body in cells obtained by **amniocentesis** or **chorionic villus sampling** used to be used as an indication of the sex of a baby before birth.

Base pair (bp)
Two **nucleotides** held together by hydrogen bonds. In **DNA**, **guanine** always pairs with **cytosine**, and **thymine** with **adenine**. A base pair is also the basic unit for measuring DNA length.

C

Carrier An individual who is **heterozygous** for a mutant **allele** (ie, carries one wild-type [normal copy] and one mutated copy of the **gene** under consideration).

CentiMorgan (cM) Unit of genetic distance. If the chance of **recombination** between two loci is 1%, the loci are said to be 1 cM apart. On average, 1 cM implies a physical distance of 1 Mb (1,000,000 **base pairs**) but significant deviations from this rule of thumb occur because recombination frequencies vary throughout the **genome**. Thus if recombination in a certain region is less likely than average, 1 cM may be equivalent to 5 Mb (5,000,000 base pairs) in that region.

Centromere Central constriction of the **chromosome** where daughter **chromatids** are joined together, separating the short (p) from the long (q) arms (see **Figure 4**).

Chorionic villus sampling (CVS) Prenatal diagnostic procedure for obtaining fetal tissue at an earlier stage of gestation than **amniocentesis**. Generally performed after 10 weeks, ultrasound is used to guide aspiration of tissue from the villus area of the chorion.

Chromatid One of the two parallel identical strands of a **chromosome**, connected at the **centromere** during **mitosis** and **meiosis** (see **Figure 4**). Before replication, each chromosome consists of only one chromatid. After replication, two identical sister chromatids are present. At the end of mitosis or meiosis, the two sisters separate and move to opposite poles before the cell splits.

Chromatin A readily stained substance in the nucleus of a cell consisting of **DNA** and proteins. During cell division it coils and folds to form the metaphase **chromosomes**.

Chromosome One of the threadlike "packages" of **genes** and other **DNA** in the nucleus of a cell (see **Figure 4**). Humans have 23 pairs of chromosomes, 46 in total: 44 **autosomes** and two **sex chromosomes**. Each parent contributes one chromosome to each pair.

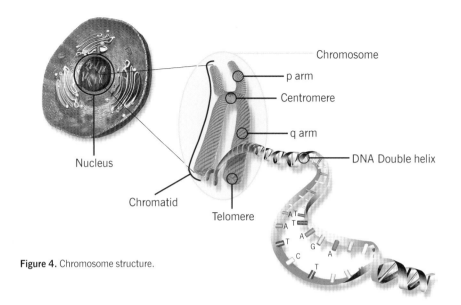

Figure 4. Chromosome structure.

Chromosomal disorder	A disorder that results from gross changes in **chromosome** dose. May result from addition or loss of entire chromosomes or just portions of chromosomes.
Clone	A group of genetically identical cells with a common ancestor.
Codon	A 3-base coding unit of **DNA** that specifies the function of a corresponding unit (**anticodon**) of **transfer RNA** (tRNA).
Complementary DNA (cDNA)	**DNA** synthesized from **messenger RNA** (mRNA) using **reverse transcriptase**. Differs from **genomic** DNA because it lacks **introns**.
Complementation	The wild-type **allele** of a **gene** compensates for a mutant allele of the same gene so that the heterozygote's **phenotype** is wild-type.
Complementation analysis	A genetic test (usually performed *in vitro*) that determines whether or not two **mutations** that produce the same **phenotype** are allelic. It enables the geneticist to determine how many distinct **genes** are involved when confronted with a number of mutations that have similar phenotypes.

Occasionally it can be observed clinically. Two parents who both suffer from **recessive** deafness (ie, both are **homozygous** for a mutation resulting in deafness) may have offspring that have normal hearing. If A and B refer to the wild-type (normal) forms of the genes, and a and b the mutated forms, one parent could be aa,BB and the other AA,bb. If **alleles** A and B are distinct, each child will have the **genotype** aA,bB and will have normal hearing. If A and B are allelic, the child will be homozygous at this **locus** and will also suffer from deafness.

Compound heterozygote	An individual with two different mutant **alleles** at the same **locus**.
Concordant	A pair of twins who manifest the same **phenotype** as each other.
Consanguinity	Sharing a common ancestor, and thus genetically related. **Recessive** disorders are seen with increased frequency in consanguineous families.
Consultand	An individual seeking genetic advice.
Contiguous gene syndrome	A syndrome resulting from the simultaneous functional imbalance of a group of **genes** (see **Figure 5**). The nomenclature for this group of disorders is somewhat confused, largely as a result of the history of their elucidation. The terms submicroscopic rearrangement/deletion/duplication and microrearrangement/deletion/duplication are often used interchangeably. Micro or submicroscopic refer to the fact that such lesions are not detectable with standard cytogenetic approaches (where the limit of resolution is usually 10 Mb, and 5 Mb in only the most fortuitous of circumstances). A newer, and perhaps more comprehensive, term that is currently applied to this group of disorders is segmental aneusomy syndromes (SASs). This term embraces the possibility not only of loss or gain of a chromosomal region that harbors many genes (leading to imbalance of all those genes), but also of functional imbalance in a group of genes, as a result of an abnormality of the machinery involved in their silencing/**transcription** (ie, methylation-based mechanisms that depend on a master control gene).

In practice, most contiguous gene syndromes result from the **heterozygous** deletion of a segment of **DNA** that is large in molecular terms but not detectable cytogenetically. The size of such deletions is usually 1.5–3.0 Mb. It is common for one to two dozen genes to be involved in such deletions, and the resultant **phenotypes** are often

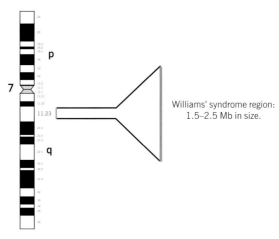

Figure 5. Schematic demonstrating the common deletion found in Williams' syndrome, at 7q11.23. The common deletion is not detectable using standard cytogenetic analysis (even high resolution), despite the fact that the deletion is at least 1.5 Mb in size. In practice, only genomic rearrangements that affect at least 5–10 Mb are detectable, either by standard cytogenetic analysis or, in fact, any technique whose endpoint involves analysis at the chromosomal level. Such deletions are termed microdeletions or submicroscopic deletions. Approximately 20 genes are known to be involved in the 7q11.23 microdeletion, and work is underway to determine which genes contribute to which aspects of the Williams' syndrome phenotype.

complex, involving multiple organ systems and, almost invariably, learning difficulties. A good example of a contiguous gene syndrome is Williams' syndrome, a sporadic disorder that is due to a heterozygous deletion at **chromosome** 7q11.23. Affected individuals have characteristic phenotypes, including recognizable facial appearance and typical behavioral traits (including moderate learning difficulties). Velocardiofacial syndrome is currently the most common **microdeletion** known, and is caused by deletions of 3 Mb at chromosome 22q11.

Crossing over   Reciprocal exchange of genetic material between **homologous chromosomes** at **meiosis** (see **Figure 6**).

Cytogenetics   The study of the structure of **chromosomes**.

Cytosine (C)   One of the bases making up **DNA** and **RNA** (pairs with **guanine**).

Cytotrophoblast   Cells obtained from fetal chorionic villi by **chorionic villus sampling (CVS)**. Used for **DNA** and **chromosome** analysis.

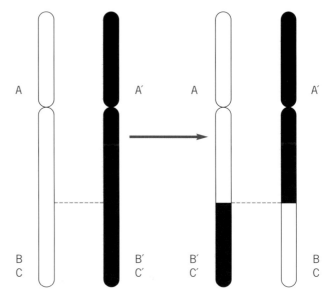

Figure 6. Schematic demonstrating the principle of **recombination (crossing over)**. On average, 50 recombinations occur per meiotic division (1–2 per **chromosome**). **Loci** that are far apart on the chromosome are more likely to be separated during recombination than those that are physically close to each other (they are said to be linked, see **linkage**), ie, A and B are less likely to cosegregate than B and C. Note that the two **homologues** of a sequence have been differentially labeled according to their chromosome of origin.

D

Deletion

A particular kind of **mutation** that involves the loss of a segment of **DNA** from a **chromosome** with subsequent re-joining of the two extant ends. It can refer to the removal of one or more bases within a **gene** or to a much larger aberration involving millions of bases. The term deletion is not totally specific, and differentiation must be made between **heterozygous** and **homozygous** deletions. Large heterozygous deletions are a common cause of complex **phenotypes** (see **contiguous gene syndrome**); large germ-line homozygous deletions are extremely rare, but have been described. Homozygous deletions are frequently described in **somatic cells**, in association with the manifestation of the malignant phenotype. The two deletions in a homozygous deletion need not be identical, but must result in the complete absence of DNA sequences that occupy the "overlap" region.

Denature	Broadly used to describe two general phenomena:
	1. The "melting" or separation of double-stranded **DNA** (dsDNA) into its constituent single strands, which may be achieved using heat or chemical approaches.
	2. The denaturation of proteins. The specificity of proteins is a result of their 3-dimensional conformation, which is a function of their (linear) amino acid sequence. Heat and/or chemical approaches may result in denaturation of a protein – the protein loses its 3-dimensional conformation (usually irreversibly) and, with it, its specific activity.
Diploid	Having two sets of chromosomes. The number of **chromosomes** in most human **somatic cells** is 46. This is double the number found in **gametes** (23, the **haploid** number).
Discordant	A pair of twins who differ in their manifestation of a **phenotype**.
Dizygotic	The fertilization of two separate eggs by two separate sperm resulting in a pair of genetically nonidentical twins.
DNA (deoxyribonucleic acid)	The molecule of heredity. DNA normally exists as a double-stranded (ds) molecule, one strand is the complement (in sequence) of the other. The two strands are joined together by hydrogen bonding, a noncovalent mechanism that is easily reversible using heat or chemical means. DNA consists of four distinct bases: **guanine** (G), **cytosine** (C), **thymine** (T), and **adenine** (A). The convention is that DNA sequences are written in a 5′ to 3′ direction, where 5′ and 3′ refer to the numbering of carbons on the deoxyribose ring. A guanine on one strand will always pair with a cytosine on the other strand, while thymine pairs with adenine. Thus, given the sequence of bases on one strand, the sequence on the other is immediately determined:
	5′–AGTGTGACTGATCTTGGTG–3′ 3′–TCACACTGACTAGAACCAC–5′
	The complexity (informational content) of a DNA molecule resides almost completely in the particular sequence of its bases. For a sequence of length "n" **base pairs**, there are 4^n possible sequences. Even for relatively small n, this number is astronomical ($4^n = 1.6 \times 10^{60}$ for n = 100).

The complementarity of the two strands of a dsDNA molecule is a very important feature and one that is exploited in almost all molecular genetic techniques. If dsDNA is **denatured**, either by heat or by chemical means, the two strands become separated from each other. If the conditions are subsequently altered (eg, by reducing heat), the two strands eventually "find" each other in solution and re-anneal to form dsDNA once again. The specificity of this reaction is quite high, under the right circumstances – strands that are not highly complementary are much less likely to re-anneal compared to perfect or near perfect matches. The process by which the two strands "find" each other depends on random molecular collisions, and a "**zippering**" mechanism, which is initiated from a short stretch of complementarity. This property of DNA is vital for the **polymerase chain reaction (PCR)**, **Southern blotting**, and any method that relies on the use of a DNA/**RNA probe** to detect its counterpart in a complex mix of molecules.

DNA chip

A "chip" or microarray of multiple **DNA** sequences immobilized on a solid surface (see **Figure 7**). The term chip refers more often to semiconductor-based DNA arrays, in which short DNA sequences (oligos) are synthesized *in situ*, using a photolithographic process akin to that used in the manufacture of semiconductor devices for the electronics industry. The term microarray is much more general and includes any collection of DNA sequences immobilized onto a solid surface, whether by a photolithographic process, or by simple "spotting" of DNA sequences onto glass slides.

The power of DNA microarrays is based on the parallel analysis that they allow for. In conventional **hybridization** analysis (ie, **Southern blotting**), a single DNA sequence is usually used to interrogate a small number of different individuals. In DNA microarray analysis, this approach is reversed – an individual's DNA is hybridized to an array that may contain 30,000 distinct spots. This allows for direct information to be obtained about all DNA sequences on the array in one experiment. DNA microarrays have been used successfully to directly uncover **point mutations** in single **genes**, as well as detect alterations in **gene expression** associated with certain disease states/cellular differentiation. It is likely that certain types of array will be useful in the determination of subtle copy number alterations, as occurs in **microdeletion**/**microduplication** syndromes.

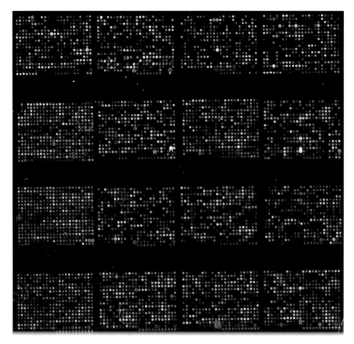

Figure 7 DNA chip. DNA arrays (or "chips") are composed of thousands of "spots" of DNA, attached to a solid surface (normally glass). Each spot contains a different DNA sequence. The arrays allow for massively parallel experiments to be performed on samples. In practice, two samples are applied to the array. One sample is a control (from a "normal" sample) and one is the test sample. Each sample is labeled with fluorescent tags, control with green and test with red. The two labeled samples are cohybridized to the array and the results read by a laser scanner. Spots on the array whose DNA content is equally represented in the test and control samples yield equal intensities in the red and green channels, resulting in a yellow signal. Spots appearing as red represent DNA sequences that are present at higher concentration in the test sample compared to the control sample and vice versa.

DNA methylation   Addition of a methyl group ($-CH_3$) to **DNA nucleotides** (often **cytosine**). Methylation is often associated with reduced levels of expression of a given **gene** and is important in **imprinting**.

DNA replication   Use of existing **DNA** as a template for the synthesis of new DNA strands. In humans and other eukaryotes, replication takes place in the cell nucleus. DNA replication is semiconservative – each new double-stranded molecule is composed of a newly synthesized strand and a pre-existing strand.

Dominant (traits/diseases)	Manifesting a **phenotype** in the **heterozygous** state. Individuals with Huntington's disease, a dominant condition, are affected even though they possess one normal copy of the **gene**.
Dynamic/ nonstable mutation	The vast majority of **mutations** known to be associated with human genetic disease are intergenerationally stable (no alteration in the mutation is observed when transmitted from parent to child). However, a recently described and growing class of disorders result from the presence of mutations that are unstable intergenerationally. These disorders result from the presence of tandem repeats of short **DNA** sequences (eg, the sequence CAG may be repeated many times in tandem), see **Table 1**. For reasons that are not completely clear, the copy number of such repeats may vary from parent to child (usually resulting in a copy number increase) and within the **somatic cells** of a given individual. Abnormal **phenotypes** result when the number of repeats reaches a given threshold. Furthermore, when this threshold has been reached, the risk of even greater expansion of copy number in subsequent generations increases.

E

Electrophoresis	The separation of molecules according to size and ionic charge by an electrical current.
	Agarose gel electrophoresis Separation, based on size, of **DNA/RNA** molecules through agarose. Conventional agarose gel electrophoresis generally refers to electrophoresis carried out under standard conditions, allowing the resolution of molecules that vary in size from a few hundred to a few thousand **base pairs**.
	Polyacrylamide gel electrophoresis Allows resolution of proteins or DNA molecules differing in size by only 1 base pair.
	Pulsed field gel electrophoresis (Also performed using agarose) refers to a specialist technique that allows resolution of much larger DNA molecules, in some cases up to a few Mb in size.

Disorder	Protein/location	Repeat	Repeat location	Normal range	Pre-mutation	Full mutation	Type	MIM
Progressive myoclonus epilepsy of Unverricht-Lundborg type (EPM1)	cystatin B 21q22.3	C₄GC₄G CG	Promoter	2–3	12–17	30–75	AR	254800
Fragile X type A (FRAXA)	FMR1 Xq27.3	CGG	5'UTR	6–52	~60–200	~200–>2,000	XLR	309550
Fragile X type E (FRAXE)	FMR2 Xq28	CGG 5	C'UTR	6–25	–	>200	XLR	309548
Friedreich's ataxia (FRDA)	frataxin 9q13	GAA	intron	17–22	–	200–>900	AR	229300
Huntington's disease (HD)	huntingtin 4p16.3	CAG	ORF	6–34	–	36–180	AD	143100
Dentatorubal-pallidoluysian atrophy (DRPLA)	atrophin 12p12	CAG	ORF	7–25	–	49–88	AD	125370
Spinal and bulbar muscular atrophy (SBMA – Kennedy syndrome)	androgen receptor Xq11-12	CAG	ORF	11–24	–	40–62	XLR	313200
Spinocerebellar ataxia type 1 (SCA1)	ataxin-1 6p23	CAG	ORF	6–39	–	39–83	AD	164400
Spinocerebellar ataxia type 2 (SCA2)	ataxin-2 12q24	CAG	ORF	15–29	–	34–59	AD	183090
Spinocerebellar ataxia type 3 (SCA3)	ataxin-3 14q24.3-q31	CAG	ORF	13–36	–	55–84	AD	109150
Spinocerebellar ataxia type 6 (SCA6)	PQ calcium channel 19p13	CAG	ORF	4–16	–	21–30	AD	183086
Spinocerebellar ataxia type 7 (SCA7)	ataxin-7 3p21.1-p12	CAG	ORF	4–35	28–35	34–>300	AD	164500
Spinocerebellar ataxia type 8 (SCA8)	SCA8 13q21	CTG	3'UTR	6–37	–	~107–250[1]	AD	603680
Spinocerebellar ataxia type 10 (SCA10)	SCA10 22q13-qter	ATTCT	intron 9	10–22	–	500–4,500	AD	603516
Spinocerebellar ataxia type 12 (SCA12)	PP2R2B 5q31-33	CAG	5'UTR	7–28	–	66–78	AD	604326
Myotonic dystrophy (DM)	DMPK 19q13.3	CTG	3'UTR	5–37	~50–180	~200–>2,000	AD	160900

Table 1. "Classical" repeat expansion disorders. [1]Longer alleles exist but are not associated with disease. AD: autosomal dominant; AR: autosomal recessive; ORF: open reading frame (coding region); 3′ UTR: 3′ untranslated region (downstream of gene); 5′ UTR: 5′ untranslated region (upstream of gene); XLR: X-linked recessive.

Empirical recurrence risk – recurrence risk   Based on observation, rather than detailed knowledge of, eg, modes of inheritance or environmental factors.

Endonuclease   An enzyme that cleaves **DNA** at an internal site (see also **restriction enzyme**).

Euchromatin   **Chromatin** that stains lightly with trypsin G banding and contains active/potentially active **genes**.

Euploidy   Having a normal **chromosome** complement.

Exon	Coding part of a **gene**. Historically, it was believed that all of a **DNA** sequence is mirrored exactly on the messenger **RNA** (mRNA) molecule (except for the presence of **uracil** in mRNA compared to **thymine** in DNA). It was a surprise to discover that this is generally not the case. The **genomic** sequence of a gene has two components: **exons** and **introns**. The exons are found in both the genomic sequence and the mRNA, whereas the introns are found only in the genomic sequence. The mRNA for dystrophin, an **X-linked** gene associated with Duchenne muscular dystrophy (DMD), is 14,000 **base pairs** long but the genomic sequence is spread over a distance of 1.5 million base pairs, because of the presence of very long intronic sequences. After the genomic sequence is initially transcribed to RNA, a complex system ensures specific removal of introns. This system is known as **splicing**.
Expressivity	Degree of expression of a disease. In some disorders, individuals carrying the same **mutation** may manifest wide variability in severity of the disorder. **Autosomal dominant** disorders are often associated with **variable expressivity**, a good example being Marfan's syndrome. Variable expressivity is to be differentiated from **incomplete penetrance**, an all or none phenomenon that refers to the complete absence of a **phenotype** in some **obligate carriers**.

F

Familial	Any trait that has a higher frequency in relatives of an affected individual than the general population.
FISH	Fluorescence *in situ* hybridization (see ***In situ* hybridization**).
Founder effect	The high frequency of a mutant **allele** in a population as a result of its presence in a founder (ancestor). Founder effects are particularly noticeable in relative genetic isolates, such as the Finnish or Amish.
Frame-shift mutation	**Deletion/insertion** of a **DNA** sequence that is not an exact multiple of 3 **base pairs**. The result is an alteration of the reading frame of the **gene** such that all sequence that lies beyond the **mutation** is missence (ie, codes for the wrong amino acids) (see **Figure 8**). A premature **stop codon** is usually encountered shortly after the frame shift.

Genetics for ENT Specialists

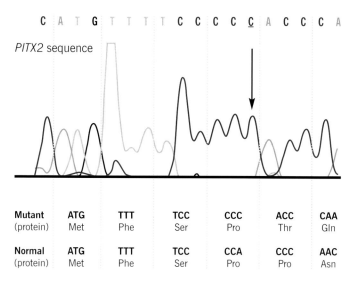

Mutant (protein)	ATG Met	TTT Phe	TCC Ser	CCC Pro	ACC Thr	CAA Gln
Normal (protein)	ATG Met	TTT Phe	TCC Ser	CCA Pro	CCC Pro	AAC Asn

Figure 8. Frame-shift mutation. This example shows a sequence of *PITX2* in a patient with Rieger's syndrome, an **autosomal dominant** condition. The sequence graph shows only the abnormal sequence. The arrow indicates the insertion of a single **cytosine** (C) residue. When translated, the triplet code is now out of frame by 1 base pair. This totally alters the translated protein's amino acid sequence. This leads to a premature **stop codon** later in the protein and results in Rieger's syndrome.

G

Gamete (germ cell)	The mature male or female reproductive cells, which contain a **haploid** set of **chromosomes**.
Gene	An ordered, specific sequence of **nucleotides** that controls the transmission and expression of one or more traits by specifying the sequence and structure of a particular protein or **RNA** molecule. Mendel defined a gene as the basic physical and functional unit of all heredity.
Gene expression	The process of converting a **gene's** coded information into the existing, operating structures in the cell.
Gene mapping	Determines the relative positions of **genes** on a **DNA** molecule and plots the genetic distance in **linkage** units (**centiMorgans**) or physical distance (**base pairs**) between them.
Genetic code	Relationship between the sequence of bases in a nucleic acid and the order of amino acids in the polypeptide synthesized from it

		2nd	2nd	2nd	2nd		
		T	C	A	G		3rd
1st	T	TTT Phe [F]	TCT Ser [S]	TAT Tyr [Y]	TGT Cys [C]	T	3rd
		TTC Phe [F]	TCC Ser [S]	TAC Tyr [Y]	TGC Cys [C]	C	
		TTA Leu [L]	TCA Ser [S]	TAA Ter [end]	TGA Ter [end]	A	
		TTG Leu [L]	TCG Ser [S]	**TAG Ter [end]**	TGG Trp [W]	G	
1st	C	CTT Leu [L]	CCT Pro [P]	CAT His [H]	CGT Arg [R]	T	3rd
		CTC Leu [L]	CCC Pro [P]	CAC His [H]	CGC Arg [R]	C	
		CTA Leu [L]	CCA Pro [P]	CAA Gln [Q]	CGA Arg [R]	A	
		CTG Leu [L]	CCG Pro [P]	CAG Gln [Q]	CGG Arg [R]	G	
1st	A	ATT Ile [I]	ACT Thr [T]	AAT Asn [N]	AGT Ser [S]	T	3rd
		ATC Ile [I]	ACC Thr [T]	AAC Asn [N]	AGC Ser [S]	C	
		ATA Ile [I]	ACA Thr [T]	AAA Lys [K]	AGA Arg [R]	A	
		ATG Met [M]	ACG Thr [T]	AAG Lys [K]	AGG Arg [R]	G	
1st	G	GTT Val [V]	GCT Ala [A]	GAT Asp [D]	GGT Gly [G]	T	3rd
		GTC Val [V]	GCC Ala [A]	GAC Asp [D]	GGC Gly [G]	C	
		GTA Val [V]	GCA Ala [A]	GAA Glu [E]	GGA Gly [G]	A	
		GTG Val [V]	GCG Ala [A]	GAG Glu [E]	GGG Gly [G]	G	

Table 2. The genetic code. To locate a particular codon (eg, TAG, marked in bold) locate the first base (T) in the left hand column, then the second base (A) by looking at the top row, and finally the third (G) in the right hand column (TAG is a stop codon). Note the redundancy of the genetic code – for example, three different codons specify a stop signal, and threonine (Thr) is specified by any of ACT, ACC, ACA, and ACG.

(see **Table 2**). A sequence of three nucleic acid bases (a triplet) acts as a codeword (**codon**) for one amino acid or instruction (start/stop).

Genetic counseling Information/advice given to families with, or at risk of, genetic disease. Genetic counseling is a complex discipline that requires accurate diagnostic approaches, up-to-date knowledge of the genetics of the condition, an insight into the beliefs/anxieties/wishes of the individual seeking advice, intelligent risk estimation, and, above all, skill in communicating relevant information to individuals from a wide variety of educational backgrounds. Genetic counseling is most often carried out by trained medical geneticists or, in some countries, specialist genetic counselors or nurses.

Genetic heterogeneity Association of a specific **phenotype** with **mutations** at different loci. The broader the phenotypic criteria, the greater the heterogeneity (eg, mental retardation). However, even very specific phenotypes may be genetically heterogeneous. Isolated central hypothyroidism is a good example: this **autosomal recessive** condition is now known to be associated (in different individuals) with mutations in the TSH β chain at 1p13, the TRH receptor at 8q23, or TRH itself at 3q13.3–q21. There is no obvious distinction between the clinical phenotypes associated with these two genes. Genetic heterogeneity should not be confused with **allelic heterogeneity**, which refers to the presence of different mutations at the same **locus**.

Genetic locus   A specific location on a **chromosome**.

Genetic map   A map of genetic landmarks deduced from **linkage (recombination) analysis**. Aims to determine the linear order of a set of **genetic markers** along a **chromosome**. Genetic maps differ significantly from **physical maps**, in that recombination frequencies are not identical across different **genomic** regions, resulting occasionally in large discrepancies.

Genetic marker   A **gene** that has an easily identifiable **phenotype** so that one can distinguish between those cells or individuals that do or do not have the gene. Such a gene can also be used as a **probe** to mark cell nuclei or **chromosomes**, so that they can be isolated easily or identified from other nuclei or chromosomes later.

Genetic screening   Population analysis designed to ascertain individuals at risk of either suffering or transmitting a genetic disease.

Genetically lethal   Preventing reproduction of the individual, either by causing death prior to reproductive age, or as a result of social factors making it highly unlikely (although not impossible) that the individual concerned will reproduce.

Genome   The complete **DNA** sequence of an individual, including the **sex chromosomes** and **mitochondrial DNA** (mtDNA). The genome of humans is estimated to have a complexity of 3.3×10^9 **base pairs** (per **haploid** genome).

Genomic   Pertaining to the **genome**. Genomic **DNA** differs from **complementary DNA** (cDNA) in that it contains noncoding as well as coding DNA.

Genotype   Genetic constitution of an individual, distinct from expressed features (**phenotype**).

Germ line   Germ cells (those cells that produce **haploid gametes**) and the cells from which they arise. The germ line is formed very early in embryonic development. Germ line **mutations** are those present constitutionally in an individual (ie, in all cells of the body) as opposed to somatic mutations, which affect only a proportion of cells.

Giemsa banding	Light/dark bar code obtained by staining **chromosomes** with Giemsa stain. Results in a unique bar code for each chromosome.
Guanine (G)	One of the bases making up **DNA** and **RNA** (pairs with **cytosine**).

H

Haploid	The **chromosome** number of a normal **gamete**, containing one each of every individual chromosome (23 in humans).
Haploinsufficiency	The presence of one active copy of a **gene**/region is insufficient to compensate for the absence of the other copy. Most genes are not "haploinsufficient" – 50% reduction of gene activity does not lead to an abnormal **phenotype**. However, for some genes, most often those involved in early development, reduction to 50% often correlates with an abnormal phenotype. Haploinsufficiency is an important component of most **contiguous gene disorders** (eg, in Williams' syndrome, **heterozygous deletion** of a number of genes results in the mutant phenotype, despite the presence of normal copies of all affected genes).
Hemizygous	Having only one copy of a **gene** or **DNA** sequence in **diploid** cells. Males are hemizygous for most genes on the **sex chromosomes**, as they possess only one **X chromosome** and one **Y chromosome** (the exceptions being those genes with counterparts on both sex chromosomes). **Deletions** on **autosomes** produce hemizygosity in both males and females.
Heterochromatin	Contains few active **genes**, but is rich in highly repeated simple sequence **DNA**, sometimes known as satellite DNA. Heterochromatin refers to inactive regions of the **genome**, as opposed to **euchromatin**, which refers to active, gene expressing regions. Heterochromatin stains darkly with **Giemsa**.
Heterozygous	Presence of two different **alleles** at a given **locus**.
Histones	Simple proteins bound to **DNA** in **chromosomes**. They help to maintain **chromatin** structure and play an important role in regulating **gene** expression.

Holandric	Pattern of inheritance displayed by **mutations** in **genes** located only on the **Y chromosome**. Such mutations are transmitted only from father to son.
Homologue or homologous gene	Two or more **genes** whose sequences manifest significant similarity because of a close evolutionary relationship. May be between species (orthologues) or within a species (paralogues).
Homologous chromosomes	**Chromosomes** that pair during **meiosis**. These chromosomes contain the same linear **gene** sequences as one another and derive from one parent.
Homology	Similarity in **DNA** or protein sequences between individuals of the same species or among different species.
Homozygous	Presence of identical **alleles** at a given **locus**.
Human gene therapy	The study of approaches to treatment of human genetic disease, using the methods of modern molecular genetics. Many trials are under way studying a variety of disorders, including cystic fibrosis. Some disorders are likely to be more treatable than others – it is probably going to be easier to replace defective or absent **gene** sequences rather than deal with genes whose aberrant expression results in an actively toxic effect.
Human genome project	Worldwide collaboration aimed at obtaining a complete sequence of the human **genome**. Most sequencing has been carried out in the USA, although the Sanger Centre in Cambridge, UK has sequenced one third of the genome, and centers in Japan and Europe have also contributed significantly. The first draft of the human genome was released in the summer of 2000 to much acclaim. Celera, a privately funded venture, headed by Dr Craig Ventner, also published its first draft at the same time.
Hybridization	Pairing of complementary strands of nucleic acid. Also known as **re-annealing**. May refer to re-annealing of **DNA** in solution, on a membrane (**Southern blotting**) or on a DNA microarray. May also be used to refer to fusion of two **somatic cells**, resulting in a hybrid that contains genetic information from both donors.

I

Imprinting A general term used to describe the phenomenon whereby a **DNA** sequence (coding or otherwise) carries a signal or imprint that indicates its parent of origin. For most DNA sequences, no distinction can be made between those arising paternally and those arising maternally (apart from subtle sequence variations); for imprinted sequences this is not the case. The mechanistic basis of imprinting is almost always **methylation** – for certain **genes**, the copy that has been inherited from the father is methylated, while the maternal copy is not. The situation may be reversed for other imprinted genes. Note that imprinting of a gene refers to the general phenomenon, not which parental copy is methylated (and, therefore, usually inactive). Thus, formally speaking, it is incorrect to say that a gene undergoes paternal imprinting. It is correct to say that the gene undergoes imprinting and that the inactive (methylated) copy is always the paternal one. However, in common genetics parlance, paternal imprinting is usually understood to mean the same thing.

In situ hybridization (ISH) Annealing of **DNA** sequences to immobilized **chromosomes**/cells/tissues. Historically done using radioactively labeled **probes**, this is currently most often performed with fluorescently tagged molecules (fluorescent *in situ* hybridization – **FISH**, see **Figure 9**). **ISH/FISH** allows for the rapid detection of a DNA sequence within the **genome**.

Incomplete penetrance Complete absence of expression of the abnormal **phenotype** in a proportion of individuals known to be **obligate carriers**. To be distinguished from **variable expressivity**, in which the phenotype always manifests in obligate carriers, but with widely varying degrees of severity.

Index case – proband The individual through which a family medically comes to light. For example, the index case may be a baby with Down's syndrome. Can be termed propositus (if male) or proposita (if female).

Insertion Interruption of a chromosomal sequence as a result of insertion of material from elsewhere in the **genome** (either a different **chromosome**, or elsewhere from the same chromosome). Such insertions may result in abnormal **phenotypes** either because of direct interruption of a **gene** (uncommon), or because of the resulting imbalance (ie, increased dosage) when the chromosomes that contain the normal counterparts of the inserted sequence are also present.

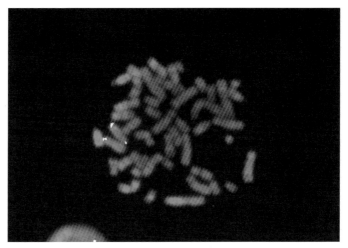

Figure 9. Fluorescence *in situ* hybridization. FISH analysis of a patient with a complex syndrome, using a clone containing **DNA** from the region 8q24.3. In addition to that clone, a control from 8pter was used. The 8pter clone has yielded a signal on both **homologues** of **chromosome** 8, while the "test" clone from 8q24.3 has yielded a signal on only one homologue, demonstrating a (**heterozygous**) deletion in that region.

Intron

A noncoding **DNA** sequence that "interrupts" the protein coding sequences of a **gene**; intron sequences are transcribed into **messenger RNA** (mRNA), but are cut out before the mRNA is translated into a protein (this process is known as **splicing**). Introns may contain sequences involved in regulating expression of a gene. Unlike the **exon**, the intron is the **nucleotide** sequence in a gene that is not represented in the amino acid sequence of the final gene product.

Inversion

A structural abnormality of a **chromosome** in which a segment is reversed, as compared to the normal orientation of the segment. An inversion may result in the reversal of a segment that lies entirely on one chromosome arm (paracentric) or one that spans (ie, contains) the **centromere** (pericentric). While individuals who possess an inversion are likely to be genetically balanced (and therefore usually phenotypically normal), they are at increased risk of producing unbalanced offspring because of problems at **meiosis** with pairing of the inversion chromosome with its normal **homologue**. Both **deletions** and duplications may result, with concomitant congenital abnormalities related to **genomic** imbalance, or miscarriage if the imbalance is lethal.

K

Karyotype

A photomicrograph of an individual's **chromosomes** arranged in a standard format showing the number, size, and shape of each chromosome type, and any abnormalities of chromosome number or morphology (see **Figure 10**).

Kilobase (kb)

1000 **base pairs** of **DNA**.

Knudson hypothesis

See **tumor suppressor gene**

L

Linkage

Coinheritance of **DNA** sequences/**phenotypes** as a result of physical proximity on a **chromosome**. Before the advent of molecular genetics, linkage was often studied with regard to proteins, enzymes, or cellular characteristics. An early study demonstrated linkage between the Duffy blood group and a form of **autosomal dominant** congenital cataract (both are now known to reside at 1q21.1). Phenotypes may also be linked in this manner (ie, families manifesting two distinct Mendelian disorders).

During the **recombination** phase of **meiosis**, genetic material is exchanged (equally) between two **homologous chromosomes**. **Genes/** DNA sequences that are located physically close to each other are unlikely to be separated during recombination. Sequences that lie far apart on the same chromosome are more likely to be separated. For sequences that reside on different chromosomes, segregation will always be random, so that there will be a 50% chance of two markers being coinherited.

Linkage analysis

An algorithm designed to map (ie, physically locate) an unknown **gene** (associated with the **phenotype** of interest) to a chromosomal region. Linkage analysis has been the mainstay of disease-associated gene identification for some years. The general availability of large numbers of DNA markers that are variable in the population (**polymorphisms**), and which therefore permit **allele** discrimination, has made linkage analysis a relatively rapid and dependable approach (see **Figure 11**). However, the method relies on the ascertainment of large families manifesting Mendelian disorders. Relatively little phenotypic

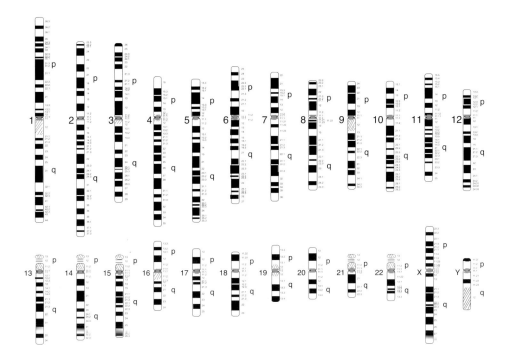

Figure 10. Schematic of a normal human (male) **karyotype**. (ISCN 550 ideogram produced by the MRC Human Genetics Unit, Edinburgh, reproduced with permission.)

heterogeneity is tolerated, as a single misassigned individual (believed to be unaffected despite being a gene **carrier**) in a **pedigree** may completely invalidate the results. **Genetic heterogeneity** is another problem, not within families (usually) but between families. Thus, conditions that result in identical phenotypes despite being associated with **mutations** within different genes (eg, tuberous sclerosis) are often hard to study. Linkage analysis typically follows a standard algorithm:

1. Large families with a given disorder are ascertained. Detailed clinical evaluation results in assignment of affected vs. unaffected individuals.

2. Large numbers of polymorphic DNA markers that span the **genome** are analyzed in all individuals (affected and unaffected).

3. The results are analyzed statistically, in the hope that one of the markers used will have demonstrably been coinherited with the phenotype in question more often than would be predicted by chance.

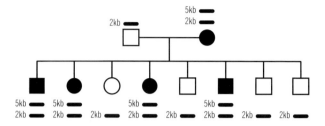

In the example above, note that the (affected) mother has a 5-kb band in addition to a 2-kb band. All the unaffected individuals have the small band only, all those who are affected have the large band. The unaffected individuals must have the mother's 2-kb fragment rather than her 5-kb fragment, and the affected individuals must have inherited the 5-kb band from the mother (as the father does not have one) – note that those individuals who only show the 2-kb band still have two alleles (one from each parent), they are just the same size and so cannot be differentiated. Thus, it appears that the 5-kb band is segregating with the disorder. The results in a family such as this are suggestive but further similar results in other families would be required for a sufficiently high **LOD** score.

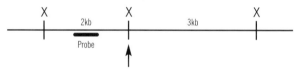

The **probe** recognizes a **DNA** sequence adjacent to a restriction site (see arrow) that is polymorphic (present on some **chromosomes** but not others). When such a site is present, the **DNA** is cleaved at that point and the probe detects a 2-kb fragment. When absent, the DNA is not cleaved and the probe detects a fragment of size (2 + 3) kb = 5 kb. X refers to the points at which the **restriction enzyme** will cleave the DNA. The recognition sequence for most restriction enzymes is very stringent – change in just one **nucleotide** will result in failure of cleavage. Most RFLPs result from the presence of a single nucleotide polymorphism that has altered the restriction site.

Figure 11. Schematic demonstrating the use of **restriction fragment length polymorphisms** (RFLPs) in **linkage analysis.**

The LOD score (**logarithm of the odds**) gives an indication of the likelihood of the result being significant (and not having occurred simply as a result of chance coinheritance of the given marker with the condition).

Linkage disequilibrium   Association of particular **DNA** sequences with each other, more often than is likely by chance alone (see **Figure 12**). Of particular relevance to inbred populations (eg, Finland), where specific disease **mutations** are found to reside in close proximity to specific variants of DNA markers, as a result of the **founder effect.**

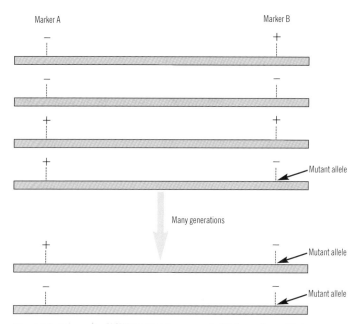

Figure 12. Schematic demonstrating the concept of **linkage disequilibrium**.

A **gene** is physically very close to marker B and further from marker A. Markers A and B, both on the same **chromosome**, can exist in one of two forms : +/−. Thus there are four possible **haplotypes**, as shown. If the **founder** mutation in the gene occurred as shown, then it is likely that even after many generations the mutant **allele** will segregate with the − form of marker B, as **recombination** is unlikely to have occurred between the two. However, since marker A is further away, the gene will now often segregate with the − form of marker A, which was not present on the original chromosome. The likelihood of recombination between the gene and marker A will depend on the physical distance between them, and on rates of recombination. It is possible that the gene would show a lesser but still significant degree of linkage disequilibrium with marker A.

Linkage map A map of **genetic markers** as determined by genetic analysis (ie, **recombination** analysis). May differ markedly from a map determined by actual physical relationships of genetic markers, because of the variability of recombination.

Locus The position of a **gene/DNA** sequence on the **genetic map**. Allelic genes/sequences are situated at identical loci in **homologous chromosomes**.

Locus heterogeneity **Mutations** at different loci cause similar **phenotypes**.

LOD (Logarithm of the Odds) score	A statistical test of **linkage**. Used to determine whether a result is likely to have occurred by chance or to truly reflect linkage. The LOD score is the logarithm (base 10) of the likelihood that the linkage is meaningful. A LOD score of 3 implies that there is only a 1:1,000 chance that the results have occurred by chance (ie, the result would be likely to occur once by chance in 1,000 simultaneous studies addressing the same question). This is taken as proof of linkage (see **Figure 11**).
Lyonization	The inactivation of n−1 **X chromosomes** on a random basis in an individual with n X chromosomes. Named after Mary Lyon, this mechanism ensures dosage compensation of **genes** encoded by the X chromosome. X chromosome inactivation does not occur in normal males who possess only one X chromosome, but does occur in one of the two X chromosomes of normal females. In males who possess more than one X chromosome (ie, XXY, XXXY, etc.), the rule is the same and only one X chromosome remains active. X-inactivation occurs in early embryonic development and is random in each cell. The inactivation pattern in each cell is faithfully maintained in all daughter cells. Therefore, females are genetic **mosaics**, in that they possess two populations of cells with respect to the X chromosome: one population has one X active, while in the other population the other X is active. This is relevant to the expression of **X-linked** disease in females.

M

Meiosis	The process of cell division by which male and female **gametes** (germ cells) are produced. Meiosis has two main roles. The first is **recombination** (during meiosis I). The second is reduction division. Human beings have 46 **chromosomes**, and each is conceived as a result of the union of two germ cells; therefore, it is reasonable to suppose that each germ cell will contain only 23 chromosomes (ie, the **haploid** number). If not, then the first generation would have 92 chromosomes, the second 184, etc. Thus, at meiosis I, the number of chromosomes is reduced from 46 to 23.
Mendelian inheritance	Refers to a particular pattern of inheritance, obeying simple rules: each **somatic cell** contains two **genes** for every characteristic and each pair of genes divides independently of all other pairs at **meiosis**.

Genetics for ENT Specialists

Mendelian Inheritance in Man (MIM/OMIM)	A catalogue of human Mendelian disorders, initiated in book form by Dr Victor McKusick of Johns Hopkins Hospital in Baltimore, USA. The original catalogue (produced in the mid-1960s) listed approximately 1500 conditions. By December 1998, this number had risen to 10,000 and by November 2003 the figure had reached 14,897. With the advent of the Internet, MIM is now available as an online resource, free of charge (OMIM – Online Mendelian Inheritance in Man). The URL for this site is: http://www.ncbi.nlm.nih.gov/omim/. The online version is updated frequently, far faster than is possible for the print version; therefore, new **gene** discoveries are quickly assimilated into the database. OMIM lists disorders according to their mode of inheritance:

1 ---- (100000–) **Autosomal dominant** (entries created before May 15, 1994)

2 ---- (200000–) **Autosomal recessive** (entries created before May 15, 1994)

3 ---- (300000–) **X-linked** loci or **phenotypes**

4 ---- (400000–) Y-linked loci or phenotypes

5 ---- (500000–) Mitochondrial loci or phenotypes

6 ---- (600000–) Autosomal loci/phenotypes (entries created after May 15, 1994).

Full explanations of the best way to search the catalogue are available at the home page for OMIM.

Messenger RNA (mRNA)	The template for protein synthesis, carries genetic information from the nucleus to the ribosomes where the code is translated into protein. Genetic information flows: **DNA** → **RNA** → protein.
Methylation	See **DNA methylation.**
Microdeletion	Structural **chromosome** abnormality involving the loss of a segment that is not detectable using conventional (even high resolution) cytogenetic analysis. Microdeletions usually involve 1–3 Mb of sequence (the resolution of cytogenetic analysis rarely is better than 10 Mb). Most microdeletions are **heterozygous**, although some individuals/families have been described with **homozygous** microdeletions. See also **contiguous gene syndrome**.

Microduplication	Structural **chromosome** abnormality involving the gain of a segment that may involve long sequences (commonly 1–3 Mb), which are, nevertheless, undetectable using conventional cytogenetic analysis. Patients with microduplications have three copies of all sequences within the duplicated segment, as compared to two copies in normal individuals. See also **contiguous gene syndrome**.
Microsatellites	DNA sequences composed of short tandem repeats (STRs), such as di- and trinucleotide repeats, distributed widely throughout the **genome** with varying numbers of copies of the repeating units. Microsatellites are very valuable as **genetic markers** for mapping human **genes**.
Missense mutation	Single base substitution resulting in a **codon** that specifies a different amino acid than the wild-type.
Mitochondrial disease/disorder	Ambiguous term referring to disorders resulting from abnormalities of mitochondrial function. Two separate possibilities should be considered.

1. **Mutations** in the mitochondrial **genome** (see **Figure 13**). Such disorders will manifest an inheritance pattern that mirrors the manner in which mitochondria are inherited. Therefore, a mother will transmit a mitochondrial mutation to all her offspring (all of whom will be affected, albeit to a variable degree). A father will not transmit the disorder to any of his offspring.

2. **Mutations** in nuclear encoded **genes** that adversely affect mitochondrial function. The mitochondrial genome does not code for all the genes required for its maintenance; many are encoded in the nuclear genome. However, the inheritance patterns will differ markedly from the category described in the first option, and will be indistinguishable from standard Mendelian disorders.

Each mitochondrion possesses between 2–10 copies of its genome, and there are approximately 100 mitochondria in each cell. Therefore, each cell possesses 200–1,000 copies of the mitochondrial genome. Heteroplasmy refers to the variability in sequence of this large number of genomes – even individuals with mitochondrial genome mutations are likely to have wild-type **alleles**. Variability in the proportion of molecules that are wild-type may have some bearing on the clinical variability often seen in such disorders.

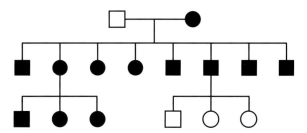

Figure 13. Mitochondrial inheritance. This **pedigree** relates to mutations in the mitochondrial **genome**.

Mitochondrial DNA	The **DNA** in the circular **chromosome** of mitochondria. Mitochondrial DNA is present in multiple copies per cell and mutates more rapidly than **genomic** (nuclear) DNA.
Mitosis	Cell division occurring in **somatic cells**, resulting in two daughter cells that are genetically identical to the parent cell.
Monogenic trait	Causally associated with a single **gene**.
Monosomy	Absence of one of a pair of **chromosomes**.
Monozygotic	Arising from a single **zygote** or fertilized egg. Monozygotic twins are genetically identical.
Mosaicism or mosaic	Refers to the presence of two or more distinct cell lines, all derived from the same **zygote**. Such cell lines differ from each other as a result of **DNA** content/sequence. Mosaicism arises when the genetic alteration occurs postfertilization (postzygotic). The important features that need to be considered in mosaicism are:

The proportion of cells that are "abnormal". In general, the greater the proportion of cells that are abnormal, the greater the severity of the associated **phenotype**.

The specific tissues that contain high levels of the abnormal cell line(s). This variable will clearly also be relevant to the manifestation of any phenotype. An individual may have a **mutation** bearing cell line in a tissue where the mutation is largely irrelevant to the normal functioning of that tissue, with a concomitant reduction in phenotypic sequelae.

Mosaicism may be functional, as in normal females who are mosaic for activity of the two **X chromosomes** (see **Lyonization**).

Mosaicism may occasionally be observed directly. **X-linked** skin disorders, such as incontinentia pigmenti, often manifest **mosaic** changes in the skin of a female, such that abnormal skin is observed alternately with normal skin, often in streaks (Blaschko's lines), which delineate developmental histories of cells.

Multifactorial inheritance	A type of hereditary pattern resulting from a complex interplay of genetic and environmental factors.
Mutation	Any heritable change in **DNA** sequence.

N

Nondisjunction	Failure of two **homologous chromosomes** to pull apart during **meiosis I**, or two **chromatids** of a chromosome to separate in meiosis II or **mitosis**. The result is that both are transmitted to one daughter cell, while the other daughter cell receives neither.
Nondynamic (stable) mutations	Stably inherited **mutations**, in contradistinction to **dynamic mutations**, which display variability from generation to generation. Includes all types of stable mutation (single base substitution, small **deletions/ insertions**, **microduplications**, and **microdeletions**).
Nonpenetrance	Failure of expression of a **phenotype** in the presence of the relevant **genotype**.
Nonsense mutation	A single base substitution resulting in the creation of a **stop codon** (see **Figure 14**).
Northern blot	**Hybridization** of a radiolabeled **RNA/DNA probe** to an immobilized RNA sequence. So called in order to differentiate it from **Southern blotting**, which was described first. Neither has any relationship to points on the compass. Southern blotting was named after its inventor Ed Southern
Nucleotide	A basic unit of **DNA** or **RNA** consisting of a nitrogenous base – **adenine**, **guanine**, **thymine**, or **cytosine** in DNA, and adenine, guanine, **uracil**, or cytosine in RNA. A nucleotide is composed of a phosphate molecule and a sugar molecule – deoxyribose in DNA and ribose in RNA. Many thousands or millions of nucleotides link to form a DNA or RNA molecule.

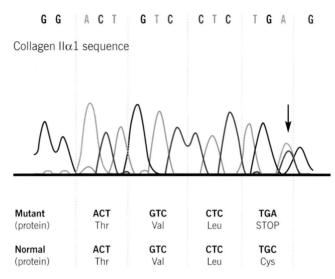

| Mutant (protein) | ACT Thr | GTC Val | CTC Leu | TGA STOP |
| Normal (protein) | ACT Thr | GTC Val | CTC Leu | TGC Cys |

Figure 14. Nonsense mutation. This example shows a sequence graph of collagen II (α1) in a patient with Stickler syndrome, an **autosomal dominant** condition. The sequence is of **genomic DNA** and shows both normal and abnormal sequences (the patient is heterozygous for the mutation).
The base marked with an arrow has been changed from G to A. When translated the codon is changed from TGC (cysteine) to TGA (stop). The premature **stop codon** in the collagen gene results in Stickler syndrome.

O

Obligate carrier	See **obligate heterozygote**.
Obligate heterozygote (obligate carrier)	An individual who, on the basis of **pedigree** analysis, must carry the mutant **allele**.
Oncogene	A **gene** that, when over expressed, causes neoplasia. This contrasts with **tumor suppressor genes**, which result in tumorigenesis when their activity is reduced.

P

p	Short arm of a **chromosome** (from the French *petit*) (see **Figure 4**).

Palindromic sequence	A **DNA** sequence that contains the same 5′ to 3′ sequence on both strands. Most **restriction enzymes** recognize palindromic sequences. An example is 5′–AGATCT–3′, which would read 3′–TCTAGA–5′ on the complementary strand. This is the recognition site of *Bgl*II.
Pedigree	A schematic for a family indicating relationships to the **proband** and how a particular disease or trait has been inherited (see **Figure 15**).
Penetrance	An all-or-none phenomenon related to the proportion of individuals with the relevant **genotype** for a disease who actually manifest the **phenotype**. Note the difference between penetrance and **variable expressivity**.
Phenotype	Observed disease/abnormality/trait. An all-embracing term that does not necessarily imply pathology. A particular phenotype may be the result of **genotype**, the environment or both.
Physical map	A map of the locations of identifiable landmarks on **DNA**, such as specific DNA sequences or **genes**, where distance is measured in **base pairs**. For any **genome**, the highest resolution map is the complete **nucleotide** sequence of the **chromosomes**. A physical map should be distinguished from a **genetic map**, which depends on **recombination** frequencies.
Plasmid	Found largely in bacterial and protozoan cells, plasmids are autonomously replicating, extrachromosomal, circular **DNA** molecules that are distinct from the normal bacterial **genome** and are often used as vectors in recombinant DNA technologies. They are not essential for cell survival under nonselective conditions, but can be incorporated into the genome and are transferred between cells if they encode a protein that would enhance survival under selective conditions (eg, an enzyme that breaks down a specific antibiotic).
Pleiotropy	Diverse effects of a single **gene** on many organ systems (eg, the **mutation** in Marfan's syndrome results in lens dislocation, aortic root dilatation, and other pathologies).
Ploidy	The number of sets of **chromosomes** in a cell. Human cells may be **haploid** (23 chromosomes, as in mature sperm or ova), **diploid** (46 chromosomes, seen in normal **somatic cells**), or triploid (69 chromosomes, seen in abnormal somatic cells, which results in severe congenital abnormalities).

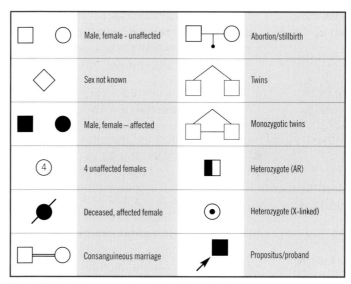

Figure 15. Symbols commonly used in **pedigree** drawing.

Point mutation   Single base substitution.

Polygenic disease   Disease (or trait) that results from the simultaneous interaction of multiple **gene** mutations, each of which contributes to the eventual **phenotype**. Generally, each **mutation** in isolation is likely to have a relatively minor effect on the phenotype. Such disorders are not inherited in a Mendelian fashion. Examples include hypertension, obesity, and diabetes.

Polymerase chain reaction (PCR)   A molecular technique for amplifying **DNA** sequences *in vitro* (see **Figure 16**). The DNA to be copied is **denatured** to its single strand form and two synthetic oligonucleotide primers are annealed to complementary regions of the target DNA in the presence of excess deoxynucleotides and a heat-stable DNA polymerase. The power of PCR lies in the exponential nature of **amplification**, which results from repeated cycling of the "copying" process. Thus, a single molecule will be copied in the first cycle, resulting in two molecules. In the second cycle, each of these will also be copied, resulting in four copies. In theory, after n cycles, there will be 2^n molecules for each starting molecule. In practice, this theoretical limit is rarely reached, mainly for technical reasons. PCR has become a standard technique in molecular biology research as well as routine diagnostics.

Polymorphism	May be applied to **phenotype** or **genotype**. The presence in a population of two or more distinct variants, such that the frequency of the rarest is at least 1% (more than can be explained by recurrent **mutation** alone). A **genetic locus** is polymorphic if its sequence exists in at least two forms in the population.
Premutation	Any **DNA mutation** that has little, if any, phenotypic consequence but predisposes future generations to the development of full mutations with phenotypic sequelae. Particularly relevant in the analysis of diseases associated with **dynamic mutations**.
Proband (propositus) – index case	The first individual to present with a disorder through which a **pedigree** can be ascertained.
Probe	General term for a molecule used to make a measurement. In molecular genetics, a probe is a piece of **DNA** or **RNA** that is labeled and used to detect its complementary sequence (eg, **Southern blotting**).
Promoter region	The noncoding sequence upstream (5′) of a **gene** where **RNA** polymerase binds. **Gene expression** is controlled by the promoter region both in terms of level and tissue specificity.
Protease	An enzyme that digests other proteins by cleaving them into small fragments. Proteases may have broad specificity or only cleave a particular site on a protein or set of proteins.
Protease inhibitor	A chemical that can inhibit the activity of a **protease**. Most proteases have a corresponding specific protease inhibitor.
Proto-oncogene	A misleading term that refers to **genes** that are usually involved in signaling and cell development, and are often expressed in actively dividing cells. Certain **mutations** in such genes may result in malignant transformation, with the mutated genes being described as **oncogenes**. The term proto-oncogene is misleading because it implies that such genes were selected for by evolution in order that, upon mutation, cancers would result because of oncogenic activation. A similar problem arises with the term **tumor suppressor gene**.
Pseudogene	Near copies of true **genes**. Pseudogenes share sequence **homology** with true genes, but are inactive as a result of multiple **mutations** over a long period of time.

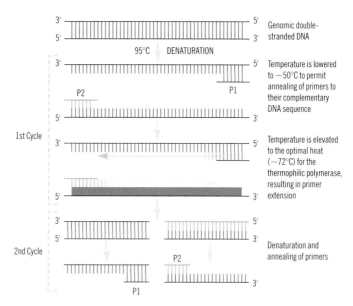

Figure 16. Schematic illustrating the technique of **polymerase chain reaction (PCR)**.

Purine	A nitrogen-containing, double-ring, basic compound occurring in nucleic acids. The purines in **DNA** and **RNA** are **adenine** and **guanine**.

Pyrimidine

A nitrogen-containing, single-ring, basic compound that occurs in nucleic acids. The pyrimidines in **DNA** are **cytosine** and **thymine**, and cytosine and **uracil** in **RNA**.

Q

q

Long arm of a **chromosome** (see **Figure 4**).

R

Re-annealing

See **hybridization**

Recessive (traits, diseases)

Manifest only in homozygotes. For the **X chromosome**, recessivity applies to males who carry only one (mutant) **allele**. Females who carry **X-linked mutations** are generally heterozygotes and, barring unfortunate X-inactivation, do not manifest X-linked recessive **phenotypes**.

Reciprocal translocation	The exchange of material between two non-**homologous chromosomes**.
Recombination	The creation of new combinations of linked **genes** as a result of **crossing over** at **meiosis** (see **Figure 6**).
Recurrence risk	The chance that a genetic disease, already present in a member of a family, will recur in that family and affect another individual.
Restriction enzyme	**Endonuclease** that cleaves double-stranded (ds)**DNA** at specific sequences. For example, the enzyme *Bgl*II recognizes the sequence AGATCT, and cleaves after the first A on both strands. Most restriction endonucleases recognize sequences that are palindromic – the complementary sequence to AGATCT, read in the same orientation, is also AGATCT. The term "restriction" refers to the function of these enzymes in nature. The organism that synthesizes a given restriction enzyme (eg, *Bgl*II) does so in order to "kill" foreign DNA – "restricting" the potential of foreign DNA that has become integrated to adversely affect the cell. The organism protects its own DNA from the restriction enzyme by simultaneously synthesizing a specific methylase that recognizes the same sequence and modifies one of the bases, such that the restriction enzyme is no longer able to cleave. Thus, for every restriction enzyme, it is likely that a corresponding methylase exists, although in practice only a relatively small number of these have been isolated.
Restriction fragment length polymorphism (RFLP)	A restriction fragment is the length of **DNA** generated when DNA is cleaved by a **restriction enzyme**. Restriction fragment length varies when a **mutation** occurs within a restriction enzyme sequence. Most commonly the **polymorphism** is a single base substitution, but it may also be a variation in length of a DNA sequence due to **variable number tandem repeats** (**VNTRs**). The analysis of the fragment lengths after DNA is cut by restriction enzymes is a valuable tool for establishing **familial** relationships and is often used in forensic analysis of blood, hair, or semen (see **Figure 11**).
Restriction map	A **DNA** sequence map, indicating the position of restriction sites.
Reverse genetics	Identification of the causative **gene** for a disorder, based purely on molecular genetic techniques, when no knowledge of the function of the gene exists (the case for most genetic disorders).

Reverse transcriptase Catalyses the synthesis of **DNA** from a single-stranded **RNA** template. Contradicted the central dogma of genetics (DNA → RNA → protein) and earned its discoverers the Nobel Prize in 1975.

RNA (ribonucleic acid) RNA molecules differ from **DNA** molecules in that they contain a ribose sugar instead of deoxyribose. There are a variety of types of RNA (including **messenger RNA**, **transfer RNA**, and ribosomal RNA) and they work together to transfer information from DNA to the protein-forming units of the cell.

Robertsonian translocation A **translocation** between two acrocentric **chromosomes**, resulting from centric fusion. The short arms and satellites (chromosome segments separated from the main body of the chromosome by a constriction and containing highly repetitive **DNA**) are lost.

S

Second hit hypothesis See **tumor suppressor gene**

Segmental aneusomy syndrome (SAS) A general term designed to encompass microdeletion/microduplication syndrome, contiguous gene syndrome, and any situation that results in loss of function of a group of genes at a particular chromosome location, irrespective of genomic copy number (ie, loss of function may be related to mutations in master control regions, which affect the expression of many genes). See also *contiguous gene syndrome*.

Sex chromosomes Refers to the **X** and **Y chromosomes**. All normal individuals possess 46 chromosomes, of which 44 are **autosomes** and two are sex chromosomes. An individual's sex is determined by his/her complement of sex chromosomes. Essentially, the presence of a Y chromosome results in the male **phenotype**. Males have an X and a Y chromosome, while females possess two X chromosomes. The Y chromosome is small and contains relatively few **genes**, concerned almost exclusively with sex determination and/or sperm formation. By contrast, the X chromosome is a large chromosome that possesses many hundreds of genes.

Sex-limited trait A trait/disorder that is almost exclusively limited to one sex and often results from **mutations** in autosomal **genes**. A good example of a sex-limited trait is breast cancer. While males are affected by breast cancer, it is much less common (~1%) than in women. Females are more prone to breast cancer than males, not only because they possess significantly more breast tissue, but also because their hormonal milieu is significantly different. In many cases, early onset bilateral breast cancer is associated with mutations either in *BRCA1* or *BRCA2*, both autosomal genes. An example of a sex-limited trait in males is male pattern baldness, which is extremely rare in premenopausal women. The inheritance of male pattern baldness is consistent with **autosomal dominant**, not **sex-linked dominant**, inheritance.

Sex-linked dominant See **X-linked dominant**

Sex-linked recessive See **X-linked recessive**

Sibship The relationship between the siblings in a family.

Silent mutation One that has no (apparent) phenotypic effect.

Single gene disorder A disorder resulting from a **mutation** on one **gene**.

Somatic cell Any cell of a multicellular organism not involved in the production of **gametes**.

Southern blot **Hybridization** with a radiolabeled **RNA/DNA probe** to an immobilized DNA sequence (see **Figure 17**). Named after Ed Southern (currently Professor of Biochemistry at Oxford University, UK), the technique has spawned the nomenclature for other types of blot (**Northern blots** for RNA and **Western blots** for proteins).

Splicing Removal of **introns** from precursor **RNA** to produce **messenger RNA** (mRNA). The process involves recognition of **intron–exon** junctions and specific removal of intronic sequences, coupled with reconnection of the two strands of **DNA** that formerly flanked the intron.

Start codon The AUG **codon** of **messenger RNA** recognized by the ribosome to begin protein production.

Stop codon The **codons** UAA, UGA, or UAG on **messenger RNA** (mRNA) (see **Table 2**). Since no **transfer RNA** (tRNA) molecules exist that possess **anticodons** to these sequences, they cannot be translated. When they occur in frame on an mRNA molecule, protein synthesis stops and the ribosome releases the mRNA and the protein.

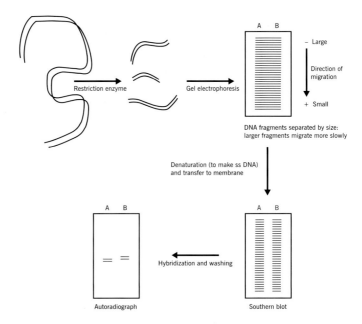

Figure 17. Southern blotting.

Synergistic heterozygosity	This refers to the phenomenon whereby the manifestation of a phenotype normally associated with complete loss of function of a single gene (ie, that gene has two mutations) may be associated with heterozygous mutations in two distinct genes that inhabit the same or related pathways.

T

Telomere	End of a **chromosome**. The telomere is a specialized structure involved in replicating and stabilizing linear **DNA** molecules.
Teratogen	Any external agent/factor that increases the probability of congenital malformations. A teratogen may be a drug, whether prescribed or illicit, or an environmental effect, such as high temperature. The classical example is thalidomide, a drug originally prescribed for morning sickness, which resulted in very high rates of congenital malformation in exposed fetuses (especially limb defects).
Termination codon	See **stop codon**.

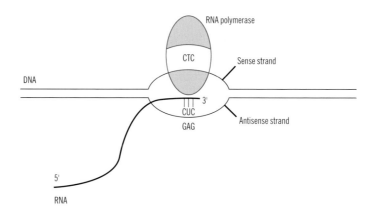

Figure 18. Schematic demonstrating the process of **transcription**. The sense strand has the sequence CTC (coding for leucine). **RNA** is generated by pairing with the antisense strand, which has the sequence GAG (the complement of CTC). The RNA produced is the complement of GAG, CUC (essentially the same as CTC, **uracil** replaces **thymine** in RNA).

Thymine (T)	One of the bases making up **DNA** and **RNA** (pairs with **adenine**).
Transcription	Synthesis of single-stranded **RNA** from a double-stranded **DNA** template (see **Figure 18**).
Transfer RNA (tRNA)	An **RNA** molecule that possesses an **anticodon** sequence (complementary to the **codon** in mRNA) and the amino acid which that codon specifies. When the ribosome "reads" the mRNA codon, the tRNA with the corresponding **anticodon** and amino acid is recruited for protein synthesis. The tRNA "gives up" its amino acid to the production of the protein.
Translation	Protein synthesis directed by a specific **messenger RNA** (mRNA), (see **Figure 19**). The information in mature mRNA is converted at the ribosome into the linear arrangement of amino acids that constitutes a protein. The mRNA consists of a series of trinucleotide sequences, known as **codons**. The **start codon** is AUG, which specifies that methionine should be inserted. For each codon, except for the **stop codons** that specify the end of translation, a **transfer RNA** (tRNA) molecule exists that possesses an **anticodon** sequence (complementary to the codon in mRNA) and the amino acid which that codon specifies. The process of translation results in the sequential addition of amino acids to the growing polypeptide chain. When translation is complete, the protein is released from the ribosome/mRNA complex and may

Genetics for ENT Specialists

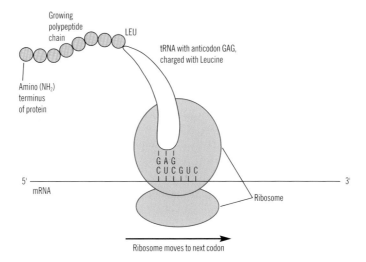

Figure 19. Schematic of the process of **translation**. **Messenger RNA** (mRNA) is translated at the ribosome into a growing polypeptide chain. For each **codon**, there is a **transfer RNA** (tRNA) molecule with the anticodon and the appropriate amino acid. Here, the amino acid leucine is shown being added to the polypeptide. The next codon is GUC, specifying valine. Translation happens in a 5′ to 3′ direction along the mRNA molecule. When the stop codon is reached, the polypeptide chain is released from the ribosome.

then undergo posttranslational modification, in addition to folding into its final, active, conformational shape.

Translocation   Exchange of chromosomal material between two or more **nonhomologous chromosomes**. Translocations may be balanced or unbalanced. Unbalanced translocations are those that are observed in association with either a loss of genetic material, a gain, or both. As with other causes of **genomic** imbalance, there are usually phenotypic consequences, in particular mental retardation. Balanced translocations are usually associated with a normal **phenotype**, but increase the risk of genomic imbalance in offspring, with expected consequences (either severe phenotypes or lethality). Translocations are described by incorporating information about the chromosomes involved (usually but not always two) and the positions on the chromosomes at which the breaks have occurred. Thus t(11;X)(p13;q27.3) refers to an apparently balanced translocation involving chromosome 11 and X, in which the break on 11 is at 11p13 and the break on the X is at Xq27.3

Triallelic inheritance   The association of a phenotype with three mutations. The classical example is Bardet–Biedl syndrome, in which some individuals only manifest the phenotype when three independent mutations are present (two on one gene and another on one of several genes implicated in this disorder). Triallelic inheritance has been trumpeted as providing an insight into the no-man's land that lies between Mendelian and polygenic disorders.

Triplet repeats   Tandem repeats in **DNA** that comprise many copies of a basic trinucleotide sequence. Of particular relevance to disorders associated with **dynamic mutations**, such as Huntington's chorea (HC). HC is associated with a pathological expansion of a CAG repeat within the coding region of the huntingtin **gene**. This repeat codes for a tract of polyglutamines in the resultant protein, and it is believed that the increase in length of the polyglutamine tract in affected individuals is toxic to cells, resulting in specific neuronal damage.

Trisomy   Possessing three copies of a particular **chromosome** instead of two.

Tumor suppressor genes   **Genes** that act to inhibit/control unrestrained growth as part of normal development. The terminology is misleading, implying that these genes function to inhibit tumor formation. The classical tumor suppressor gene is the Rb gene, which is inactivated in retinoblastoma. Unlike **oncogenes**, where a **mutation** at one **allele** is sufficient for malignant transformation in a cell (since mutations in oncogenes result in increased activity, which is unmitigated by the normal allele), both copies of a tumor suppressor gene must be inactivated in a cell for malignant transformation to proceed. Therefore, at the cellular level, tumor suppressor genes behave recessively. However, at the organismal level they behave as dominants, and an individual who possesses a mutation in only one Rb allele still has an extremely high probability of developing bilateral retinoblastomas.

The explanation for this phenomenon was first put forward by Knudson and has come to be known as the **Knudson hypothesis** (also known as the second hit hypothesis). An individual who has a germ-line mutation in one Rb allele (and the same argument may be applied to any tumor suppressor gene) will have the mutation in every cell in his/her body. It is believed that the rate of spontaneous somatic mutation (defined functionally, in terms of loss of function of that gene by whatever mechanism) is of the order of one in a million per gene per cell division.

Given that there are many more than one million retinal cells in each eye, and many cell divisions involved in retinal development, the chance that the second (wild-type) Rb allele will suffer a somatic mutation is extremely high. In a cell that has acquired a "second hit", there will now be no functional copies of the Rb gene, as the other allele is already mutated (germ-line mutation). Such a cell will have completely lost its ability to control cell growth and will eventually manifest as a retinoblastoma. The same mechanism occurs in many other tumors, the tissue affected being related to the tissue specificity of expression of the relevant tumor suppressor gene.

U

Unequal crossing over
Occurs between similar sequences on **chromosomes** that are not properly aligned. It is common where specific repeats are found and is the basis of many **microdeletion/microduplication** syndromes (see **Figure 20**).

Uniparental disomy (UPD)
In the vast majority of individuals, each **chromosome** of a pair is derived from a different parent. However, UPD occurs when an offspring receives both copies of a particular chromosome from only one of its parents. UPD of some chromosomes results in recognizable **phenotypes** whereas for other chromosomes there do not appear to be any phenotypic sequelae. One example of UPD is Prader–Willi syndrome (PWS), which can occur if an individual inherits both copies of chromosome 15 from their mother.

Uniparental heterodisomy
Uniparental disomy in which the two **homologues** inherited from the same parent are not identical. If the parent has **chromosomes** A,B the child will also have A,B.

Uniparental isodisomy
Uniparental disomy in which the two **homologues** inherited from the same parent are identical (ie, duplicates). So, if the parent has **chromosomes** A,B then the child will have either A,A or B,B.

Uracil (U)
A nitrogenous base found in **RNA** but not in **DNA**, uracil is capable of forming a **base pair** with **adenine**.

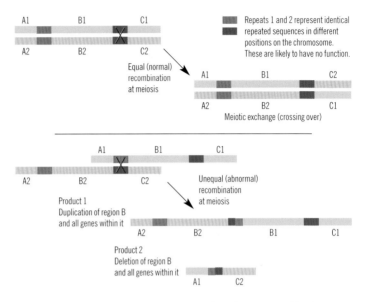

Figure 20. Schematic demonstrating (i) normal **homologous recombination** and (ii) homologous unequal recombination, resulting in a **deletion** and a duplication **chromosome**.

V

Variable expressivity Variable expression of a **phenotype**: not all-or-none (as is the case with **penetrance**). Individuals with identical **mutations** may manifest variable severity of symptoms, or symptoms that appear in one organ and not in another.

Variable number of tandem repeats (VNTR) Certain **DNA** sequences possess tandem arrays of repeated sequences. Generally, the longer the array (ie, the greater the number of copies of a given repeat), the more unstable the sequence, with a consequent wide variability between **alleles** (both within an individual and between individuals). Because of their variability, VNTRs are extremely useful for genetic studies as they allow for different alleles to be distinguished.

W

Western blot Like a **Southern** or **Northern blot** but for proteins, using a labeled antibody as a probe.

X

X-autosome translocation

Translocation between the **X chromosome** and an **autosome**.

X chromosome

See **sex chromosomes**.

X-chromosome inactivation

See **lyonization**.

X-linked

Relating to the **X chromosome**/associated with **genes** on the X chromosome.

X-linked recessive (XLR)

X-linked disorder in which the **phenotype** is manifest in **homozygous**/**hemizygous** individuals (see **Figure 21**). In practice, it is hemizygous males that are affected by X-linked recessive disorders, such as Duchenne's muscular dystrophy (DMD). Females are rarely affected by XLR disorders, although a number of mechanisms have been described that predispose females to being affected, despite being **heterozygous**.

X linked dominant (XLD)

X-linked disorder that manifests in the heterozygote. XLD disorders result in manifestation of the **phenotype** in females and males (see **Figure 22**). However, because males are **hemizygous**, they are more severely affected as a rule. In some cases, the XLD disorder results in male lethality.

Y

Y chromosome

See **sex chromosomes**.

Z

Zippering

A process by which complementary **DNA** (cDNA) strands that have annealed over a short length undergo rapid full annealing along their whole length. DNA annealing is believed to occur in two main stages. A chance encounter of two strands that are complementary results in a short region of double-stranded DNA (dsDNA), which if perfectly matched, stabilizes the two single strands so that further re-annealing

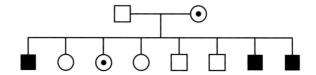

Figure 21a. X-linked recessive inheritance – A. Most X-linked disorders manifest recessively, in that **heterozygous** females (**carriers**) are unaffected and males, who are **hemizygous** (possess only one **X chromosome**) are affected. In this example, a carrier mother has transmitted the disorder to three of her sons. One of her daughters is also a carrier. On average, 50% of the male offspring of a carrier mother will be affected (having inherited the mutated X chromosome), and 50% will be unaffected. Similarly, 50% of daughters will be carriers and 50% will not be carriers. None of the female offspring will be affected but the carriers will carry the same risks to their offspring as their mother. The classical example of this type of inheritance is Duchenne muscular dystrophy.

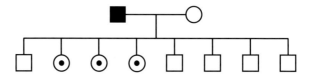

Figure 21b. X-linked recessive inheritance – B. In this example the father is affected. Because all his sons must have inherited their **Y chromosome** from him and their **X chromosome** from their normal mother, none will be affected. Since all his daughters must have inherited his X chromosome, all will be carriers but none affected. For this type of inheritance, it is clearly necessary that males reach reproductive age and are fertile – this is not the case with Duchenne's muscular dystrophy, which is usually fatal by the teenage years in boys. Emery-Dreifuss muscular dystrophy is a good example of this form of inheritance, as males are likely to live long enough to reproduce.

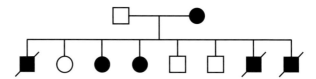

Figure 22. X-linked dominant inheritance. In X-linked dominant inheritance, the **heterozygous** female and hemizygous male are affected, however, the males are usually more severely affected than the females. In many cases, X-linked dominant disorders are lethal in males, resulting either in miscarriage or neonatal/infantile death. On average, 50% of all males of an affected mother will inherit the gene and be severely affected; 50% of males will be completely normal. Fifty percent of female offspring will have the same phenotype as their affected mother and the other 50% will be normal and carry no extra risk for their offspring. An example of this type of inheritance is incontinentia pigmenti, a disorder that is almost always lethal in males (males are usually lost during pregnancy).

Genetics for ENT Specialists

of their specific sequences proceeds extremely rapidly. The initial stage is known as nucleation, while the second stage is called zippering.

Zygote   **Diploid** cell resulting from the union of male and female **haploid gametes**.

8. Index

Cross references have been made from disease/disorder synonyms to the major terms used in this work.
Page numbers in *italics* indicate tables.
Page numbers in **bold** indicate figures.
vs indicates a comparison or differential diagnosis.

A

abductor paralysis, familial laryngeal paralysis 160
acrocephalosyndactyly
    type I *see* Apert syndrome
    type V *see* Pfeiffer syndrome
activin receptor-like kinase-1 150
adductor paralysis 162
Albers–Schönberg disease *see* osteopetrosis
ALK1 gene 150
Alport syndrome 57–60
    autosomal dominant 59–60
    autosomal recessive 59
    X-linked 57–8
anemia 96
angioneurotic edema 157–9
anosmia, Kallmann syndrome 146, 147
anosmin-1 147
Apert Crouzon disease *see* Apert syndrome
Apert syndrome 66–9
    Crouzon syndrome *vs* 68, 72
    Jackson–Weiss syndrome *vs* 68
    nonsyndromic craniosynostosis *vs* 68
    Pfeiffer syndrome *vs* 68, 74
    Saethre–Chotzen syndrome *vs* 68, 77
Arnold–Chiari malformation, Pfeiffer syndrome 74
ataxia, Leigh syndrome 138–9
ATP synthesis 128, **128**
audiograms
    DFNA1 14, **14**
    DFNA2 **17**
    DFNA5 **22**
    DFNA6/14/38 **24**
    DFNA8/12 **26**
    DFNA9 27, **28**
    DFNA10 29
    DFNA11 31, **31**
    DFNA13 32, **33**
    DFNA15 **35**

auditory brainstem (ABT) testing, Gerhardt syndrome 161
auriculobranchiogenic dysplasia *see* oculo-auriculo-vertebral spectrum
autosomal dominant Albers–Schönberg disease *see* osteopetrosis
autosomal dominant marble bones *see* osteopetrosis
autosomal dominant otospondylomegaepiphyseal dysplasia *see* Stickler
      syndrome
autosomal recessive Albers–Schönberg disease *see* osteopetrosis
autosomal recessive marble bones *see* osteopetrosis

B
BBBG syndrome *see* Opitz syndrome
BBB syndrome *see* Opitz syndrome
BCS1L gene 139
bilateral acoustic neurofibromatosis *see* neurofibromatosis, type II
bilateral acoustic schwannomas *see* neurofibromatosis, type II
bone marrow transplantation, autosomal recessive osteopetrosis 97
brachycephaly, Apert syndrome 67
brainstem-evoked responses (BER), neurofibromatosis type II 82, 83
branchial arch anomalies, branchio-oto-renal syndrome 60, **61**
branchial arch fissure, branchio-oto-renal syndrome 60, **61**
branchio-oto-renal syndrome 11, 60–3
      enlarged vestibular aqueduct syndrome *vs* 104
      inner-ear malformations 10
branchio-oto (BO) syndrome 61
branchio-renal (BR) syndrome 61
BRN3C (POU4F3) gene 5, 34
bronchitis 173
bronchodilators 173

C
C1-INH gene 158
CA2 gene 98
cadherin 23 113, 114
caloric vestibular tests, DFNA11 31
carbonic anhydrase II 98
cardiac features
      CHARGE 63, 66
      Jervell and Lange-Nielsen syndrome 78
      Kallmann syndrome, type II 146–7
      velocardiofacial syndrome 154
carotid body tumors *see* paragangliomas
cataracts 57
CDH23 gene 115–16
      DFNB12 49

Saethre–Chotzen syndrome *vs* 72, 77
cryptorchidism 146
CX26 gene *see GJB2 (CX26)* gene
CX30 gene *see GJB6 (CX30)* gene
CX31 gene *see GJB3 (CX31)* gene
cytochrome C oxidase deficiency 142

D

danazol 158
deafness *see* hearing impairment
deafness–dystonia–optic atrophy syndrome *see* Mohr Tranebjaerg
     syndrome
dentinogenesis imperfecta 40
dentin sialophosphorin (DSP) 39–40
DFN1 *see* Mohr Tranebjaerg syndrome
DFN3 51–4
DFNA1 *4*, 14–15
DFNA2 *4*, 16–19
DFNA3 *4*, 19–21
DFNA5 *4*, 21–3
DFNA5 gene *4*, 22
DFNA6/14/38 *4*, 23–5
     see also Wolfram syndrome
DFNA8/12 *4*, 25–7
DFNA9 *5*, 27–8
     Ménière's disease *vs* 27, 28
DFNA10 *5*, 29–30
DFNA11 *5*, 31–2
     see also DFNB2
DFNA13 *5*, 32–4
DFNA15 *5*, 34–5
DFNA17 *5*, 35–6
DFNA22 *5*, 36–7
     see also Usher syndrome
DFNA36 *5*, 37–8
DFNA39 *5*, 39–40
DFNA loci *4–5*, 7
DFNB1 *6*, 42–3
DFNB2 *6*, 43–4
     see also DFNA11; Usher syndrome
DFNB3 44–5
DFNB4 *6*, 103–4
     see also enlarged vestibular aqueduct syndrome (EVA)
DFNB7/DFNB11 46

DFNB8/10 *6*, 46–7
DFNB9 47–8
DFNB12 48–9
DFNB21 49–50
DFNB loci *6, 7*
diabetes–deafness syndrome, maternally transmitted 143–4
diabetes insipidus, diabetes mellitus, optic atrophy, and deafness
    syndrome *see* Wolfram syndrome
dialysis, X-linked Alport syndrome 55
DIAPH1 gene *4*, 15
DIDMOAD syndrome *see* Wolfram syndrome
digenic inheritance, DFNB1 43
diGeorge syndrome 11, 156–7
    inner-ear malformations *10*
DNA1 gene 172
DNAH5 gene 172
DNFB29 50–1
DPP (TIMM8A) gene 79
DSPP gene *5*, 39–40
Duane retraction, Wildervanck syndrome 11
dynein 172, 173
dyskinetic cilia syndrome *see* primary ciliary dyskinesis
dystonia 140
dystopia canthorum 119–20

E
ear malformations
    external *see* external ear anomalies
    inner *10*
ECGF1 gene 143
edema, angioneurotic 157–9
EDN3 gene 124
EDNRB gene 122, 124
electrocardiogram (ECG) 77
electronystagmogram (ENG) 82
electroretinogram (ERG) 115, 117
endoglin 150
endothelial cell growth factor 1 143
endothelin 3 124
endothelin receptor B 122, 124
end-stage renal disease (ESRD) 57
ENG gene 150
enlarged vestibular aqueduct syndrome (EVA) *6*, 103–4
    see also DFNB4

epibulbar dermoids 87, **88**
external ear anomalies
    CHARGE 63, **64, 65**
    Crouzon syndrome 70
    oculo-auriculo-vertebral spectrum 88
    velocardiofacial syndrome 154
EYA1 gene *10*, 62
EYA4 gene 5, 30
eye symptoms *see* ocular symptoms

F
facial features
    Apert syndrome 67, **67**
    autosomal recessive osteopetrosis 96–7
    CHARGE 63, **65**
    Crouzon syndrome 70, **71**
    enlarged vestibular aqueduct syndrome (EVA) 104
    Marshall syndrome 104, **105**
    Noonan syndrome, type I 84, **84**
    oculo-auriculo-vertebral spectrum 87, **87, 88**
    Opitz syndrome 163, **164**
    otospondylomegaepiphyseal dysplasia 108
    Pfeiffer syndrome 72–3, **73**
    Saethre–Chotzen syndrome 75, **75**
    Stickler syndrome, type I 106
    Treacher Collins' syndrome 109, **110**
    velocardiofacial syndrome 154, **155**
    Waardenburg syndrome, type I **119,** 119–20
facio-auriculo-vertebral dysplasia *see* oculo-auriculo-vertebral spectrum
familial laryngeal paralysis *see* laryngeal paralysis, familial
Fechtner syndrome 59
    MYH9 36
FGFR1 gene
    Kallmann syndrome, type II 147
    Pfeiffer syndrome 74
FGFR2 gene
    achondroplasia 72
    Apert syndrome 68
    Crouzon syndrome 71
    hypochondroplasia 72
    Pfeiffer syndrome 74
    Saethre–Chotzen syndrome 76
    thanatrophoric dwarfism 72
FGFR3 gene 76

fibroblast growth receptor 1
Kallmann syndrome, type II 147
Pfeiffer syndrome 74
fibroblast growth receptor 2
Apert syndrome 68
Crouzon syndrome 71
Pfeiffer syndrome 74
Saethre–Chotzen syndrome 76
fibroblast growth receptor 3 76
first and second brachial arch syndrome see oculo-auriculo-vertebral spectrum
first arch syndrome see oculo-auriculo-vertebral spectrum
Fisch classification 166, *166*
fractures 92, 95

G
gap-junction protein β-2
DFNA3 19
DFNB1 42
gap-junction protein β-3 18
gap-junction protein β-6 21
genitourinary abnormalities, CHARGE 63
Gerhardt syndrome 161–2
GJB2 (CX26) gene
DFNA3 19, 20
DFNB1 6, 42
GJB3 (CX31) gene
DFNA2 *4*, 18–19
DFNA3 21
GJB6 (CX30) gene
DFNA2 *4*
DFNB1 43
GL gene 97
gliomas 81
glomus jugulare tumors see paragangliomas
glomus tumors see paragangliomas
glucose tolerance test 129
goiter 101
Goldenhar–Gorlin syndrome see oculo-auriculo-vertebral spectrum
Goldenhar syndrome
Treacher Collins' syndrome *vs* 109
see *also* oculo-auriculo-vertebral spectrum
growth retardation
CHARGE 63

MNGIE syndrome 143
G syndrome *see* Opitz syndrome
Guibanud–Vainsel syndrome *see* osteopetrosis

H
harmonin 113, 114
hearing impairment
    inherited nonsyndromic 14–54
        dominant inheritance 14–41
        recessive inheritance 42–51
        X-linked 51–4
    inherited sensorineural 3–13
        age-related 3
        classification 8
        DFNA loci *4*–5, 7
        DFNB loci 6, 7
        epidemiology 3
        genetic syndromes 7–11
          anatomical sites 9
        labyrinth body abnormalities 8
        mode of inheritance 3
        nomenclature 7
        prelingual phase 3
        prevalence 3
        surgery 11–12
        X-linked 7
    inherited syndromic 54–126
    maternally transmitted *see below*
    velocardiofacial syndrome 154
    X-linked nonsyndromic 51–4
    see also specific diseases/disorders
hearing impairment, maternally transmitted
    nonsyndromic
        MTTS1-7445 130
        MTTS1-7510 132–3
        MTRNR1 gene 134–5
    oligosyndromic 131–2
Helicobacter pylori infection 158
hematuria
    autosomal recessive Alport syndrome 59
    X-linked Alport syndrome 55
hemifacial microsomia *see* oculo-auriculo-vertebral spectrum
hemignathia and microtia syndrome *see* oculo-auriculo-vertebral spectrum
hepatosplenomegaly 96

hereditary benign intraepithelial dyskeratosis, white sponge nevus *vs* 174
hereditary hemorrhagic telangiectasia *see* Rendu–Osler–Weber syndrome
heterochromia iridis 122, **122**
hydrocephalus 72
hyperostosis corticalis generalisata, osteopetrosis *vs* 100
hypertelorism
    Apert syndrome 68
    Opitz syndrome 163
hypocalcaemia 97
hypogonadism 146, 147
hypospadias 163, **164**
hypotonia 138–9

I
immotile cilia syndrome *see* primary ciliary dyskinesis
inner-ear malformations *10*
interferon 97–8
interuterine facial necrosis *see* oculo-auriculo-vertebral spectrum

J
Jackson–Weiss syndrome
    Apert syndrome *vs* 68
    Crouzon syndrome *vs* 72
    Pfeiffer syndrome *vs* 74
    Saethre–Chotzen syndrome *vs* 77
Jensen syndrome *see* Mohr Tranebjaerg syndrome
Jervell and Lange-Nielsen syndrome 77–8

K
KAL1 gene 147
Kallmann syndrome (types I–III) 146–8
Kartagener syndrome *see* primary ciliary dyskinesis
KCNE1 gene 77
KCNQ1 gene 77
KCNQ4 gene *4*, 16–17
Kearns–Sayre syndrome 135–7
keratin
    type 4 174
    type 13 174
Kippel–Feil anomaly 11
Klebattschädel skull 73
Klein–Waardenburg syndrome *see* Waardenburg syndrome, type III
KRT4 gene 174
KRT13 gene 174

kyphoscoliosis 94, 95

L
lactate levels 129
laryngeal paralysis, familial 159–62
    abductor paralysis 160
    adductor paralysis 162
    Gerhardt syndrome 161–2
    Plott syndrome 160–1
laryngotracheoesophageal abnormalities 163, **164**
lateral facial dysplasia *see* oculo-auriculo-vertebral spectrum
Leber's hereditary optic atrophy 140–1
Leber's hereditary optic neuropathy (LHON) 140–1
Leigh syndrome 138–40
leiomyomas 57
limb abnormalities
    osteogenesis imperfecta, type I **91,** 91–2
    Waardenburg syndrome, type III 123
linkage analysis, Usher syndrome 115
lipoprotein receptor-related protein 5 99
Lobstein's disease *see* osteogenesis imperfecta, type I
LRP5 gene 99

M
macrocephaly 91
macrothrombocytopenia 59
magnetic resonance imaging (MRI)
    Apert syndrome 68
    DFNA8/12 25
    enlarged vestibular aqueduct syndrome (EVA) 104
    Kallmann syndrome 147
    MERRF syndrome 142
    Mohr Tranebjaerg syndrome 79
    neurofibromatosis, type II **81,** 83
    paraganglioma, type I 168, **169**
    Pendred syndrome 103
    surgical feasibility 11–12
male Turner syndrome *see* Noonan syndrome, type I
marble bones *see* osteopetrosis
Marshall syndrome 104
maternally transmitted diabetes–deafness syndrome 143–4
May–Hegglin anomaly 36
MELAS syndrome 137–8
    MERRF syndrome *vs* 142

osteosclerosis fragilis generalisata *see* osteopetrosis
otitis media
    Apert syndrome 68
    primary ciliary dyskinesis with sinus invertus 173
    velocardiofacial syndrome 155
otocranialcephalic syndrome *see* oculo-auriculo-vertebral spectrum
otoferlin 48
OTOF gene 48
otomandibular dystosis *see* oculo-auriculo-vertebral spectrum
otosclerosis 39–41
oto-spondylo-megaepiphyseal dysplasia 108–9
    autosomal dominant *see* Stickler syndrome
    COL11A2 34
OTSC1 40
OTSC2 41
OTSC3 42
oxidative phosphorylation (OXPHOS) deficiencies 126–44
    diagnosis 128
    see also specific diseases/disorders

P

pamidronate 93, 95
palmoplantar keratoderma 20
paragangliomas 165–72
    carotid body tumors 166–7, 171
    Fisch classification 166, *166*
    Shamblin classification 166, *166*
    surgery 167
    type I 167–9
    type II 169–70
    type III 170
PAX3 gene *10*, 120, 123
PCDH15 gene 113, 114
PDHA1 gene 139
PDS (SLC26A4) gene 102
Pendred syndrome 8, 11, 101–3
    enlarged vestibular aqueduct syndrome *vs* 103
    inner-ear malformations *10*
perilymphatic gushers 51
Pfeiffer syndrome 72–4
    Apert syndrome *vs* 68, 74
    Crouzon syndrome *vs* 72, 74
PGL *see* paragangliomas
Pierre Robin syndrome

oculo-auriculo-vertebral spectrum 90
Treacher Collins' syndrome 111
see also specific types
syndactyly
Apert syndrome 67, **67**, 68
Saethre–Chotzen syndrome 75–6, **76**

T
tandem walk test 44
TCIRG1 gene 97
TCOF1 gene 109
TECTA gene
DFNA8/12 *4*, 26, 50
DFNB21 49
α-tectorin
DFNA8/12 26
DFNB21 49
telangiectases 149, **150**, 150–1
thyroidectomy 101
thyroid-stimulating hormone (TSH) 101
thyroxine 102
TIMM8A (DPP) gene 79
tinnitus
DFNA10 29
osteogenesis imperfecta, type I 92
TMC1 gene
DFNA36 5, 37
DFNB7/DFNB11 46
TMPRSS3 gene 6, 47
TNFRSF11A gene 100
Townes–Brocks syndrome 90
tracheotomy 66, 160, 161, 162
tranexamic acid 158
transforming growth factor , (TGF-,) 150–1
Treacher Collins' syndrome 109–11
Turner-like syndrome see Noonan syndrome, type I
TWIST gene 76
tympanoplasties 173

U
unilateral facial agenesis *see* oculo-auriculo-vertebral spectrum
upper airway obstruction, angioneurotic edema 157
USH1C gene 113, 114
USH2A gene 116

X

X chromosome
  Kallmann syndrome, type I 147
  Mohr Tranebjaerg syndrome 79
  Norrie's disease 86
  Opitz syndrome 164
X-linked mixed deafness with stapes fixation and perilymphatic gusher *see* DFN3
X-linked nonsyndromic hearing impediment 51–4
X-rays, osteopetrosis 97